ICLAS Manual for Genetic Monitoring of Inbred Mice

Black agouti : C3H

Black : C57BL

Brown agouti (Cinnamon) : NC

Brown (Chocolate) : C57BR

Dilute brown : DBA

Albino : AKR

Typical Coat Color of the Mice

ICLAS Manual for Genetic Monitoring of Inbred Mice

Edited by
Tatsuji Nomura
Kozaburo Esaki
Takeshi Tomita

UNIVERSITY OF TOKYO PRESS

Published by UNIVERSITY OF TOKYO PRESS, 1984
ISBN   4-13-068108-7
ISBN   0-86008-366-7

Printed in Japan.

This manual is based on the Workshop on Preparation of an International Manual on Genetical Monitoring for Inbred Strains of Mice Used in Cancer Research, held 28–30 July 1980, Tokyo, under the auspices of ICLAs (International Council for Laboratory Animal science) and the ICREW (International Cancer Research Workshop) program administered by the UICC (International Union Against Cancer).

Contents

List of Contributors

Esaki, Kozaburo   Introduction. Biochemical Markers and Data Analysis in II. Qualitative Characters in Genetic Monitoring

Festing, Michael F. W.   Maintenance and Expansion of Inbred Strains in I. Genetic Management and Embryo Banking, and III. Quantitative Characters in Genetic Monotoring

Hedrich, Hans J.   Immunogenetic Markers in II. Qualitative Characters in Genetic Monitoring

Hoffman, Harold A.   Biochemical Markers in II. Qualitative Characters in Genetic Monitoring

Mobraaten, Larry E.   Colony Management, Embryo Freezing and Banking in I. Genetic Management and Embryo Banking

Moutier, René   in Supplement

Nomura, Tatsuji   in Supplement

Radzikowski, Czesyaw   in Supplement

Tomita, Takeshi   Coat Color in II. Qualitative Characters in Genetic Monitoring

Yosida, Tosihide H.   Cytogenetic Techniques for Genetic Monitoring

Esaki, K.   Central Institute for Experimental Animals, Japan
Festing, M. F. W.*   Medical Research Council, Laboratory Animals Centre, United Kingdom
Hedrich, H. J.   Zentralinstitut für Versuchstiere, F. R. Germany
Hoffman, H. A.**   The National Institutes of Health, U.S.A.
Mobraaten, L. E.   The Jackson Laboratory, U.S.A.
Moutier, R.   Centre de Sélection et d'Elevage d'Animaux de Laboratoire, C.N.R.S., France
Nomura, T.   Central Institute for Experimental Animals, Japan
Radzikowski, Cz.   Ludwik Hirszfeld Institute of Immunology and Experimental Therapy, Poland
Tomita, T.   Faculty of Agriculture, Nagoya University, Japan
Yosida, T. H.   National Institute of Genetics, Japan

   * Present Address: Experimental Embryology and Teratology Unit
  ** Present Address: Animal Genetic Systems, Inc., North Carolina, U.S.A.

Preface

The Manual for Genetic Monitoring of Inbred Mice Used in Cancer Research was developed during a workshop organized by Dr. Tatsuji Nomura and held at the International House of Japan in Tokyo on July 28–30, 1980. The workshop was arranged under the auspices of ICLAS (International Council for Laboratory Animal Science) and the ICREW (International Cancer Research Workshop) Program administered by the International Union Against Cancer (UICC).

Following ICLAS policy and with the cooperation of former and current Governing Board Presidents and members and appropriate committee chairmen and members, the manual's goal was the introduction of standardized methods and programs for genetic monitoring of inbred strains of mice used in cancer and in other biomedical fields of research. International acceptance of the Manual may guarantee that the animals are selected by more appropriate methods, such as elimination of genetically contaminated colonies.

The participants in the Workshop represented major national laboratory animal centers and governmental, university and other research institutions concerned with the genetic quality assurance of inbred strains of mice, from the Federal Republic of Germany, France, Great Britain, Japan, Poland and the United States. During the Workshop, reports on the current status of genetic monitoring in the various institutions were presented, and in the succeeding discussions a confluence of various experiences, philosophies and approaches took place. As the result of this work, the basic aims and definitions of genetic monitoring, genetic management of inbred colonies of mice, their genetic profiles, and guidelines for the use of selected morphological, biochemical and immunological markers were determined. In addition, the importance of genetic monitoring for embryo freezing and banking was considered.

The introduction of an internationally acceptable genetic monitoring program will facilitate the exchange of information among institutions and the use of computers in genetic monitoring programs, genetic management and data banks.

This Genetic Monitoring Manual was edited on the basis of discussion and agreement at the Workshop. Special thanks are extended to our competent secretaries, Miss Yu Ohta, Mr. Nakao, and Mr. D. Havens. We are also indebted to the staff of the Central Institute for Experimental Animals for their assistance in preparing for the publication of this manual.

Tatsuji Nomura
Chairman
Committee for the ICLAS
Genetic Monitoring Manual

Introduction

Since the genetic background of laboratory animals is an important factor in the use of animals in biomedical research, many inbred strains have been established and distributed around the world. An international committee on standard genetic nomenclature for mice has recommended the definition and designation of inbred strains including substrains.

It is common belief that inbred strains are stable genetically, and that there are no genetic differences among their substrains. Indeed, the assumption of such genetic stability is a necessary prerequisite for interpreting the results of such studies with confidence. However, some random gene mutation is almost inevitable; or unexpected mating may occur across strains as the different strains are maintained from generation to generation. Several authors have reported genetic contaminations as a result of outcrossing between inbred strains and spontaneous mutations which can change the specific characteristics of inbred mouse strains. A research project on genetic quality control of laboratory animals, conducted in Japan between 1977 and 1980, brought out two serious problems. First, 4 of 10 well-known inbred mouse strains used in biomedical research had produced substrains which showed different genetic characters from those of the standard strains. Next, 8 out of 103 lines of 10 inbred strains showed segregations at two or more loci. The most likely cause of the appearance of a substrain showing unexpected genetic characters is genetic contamination in an older generation, and the most likely cause of heterozygocity is genetic contamination in recent generations.

When genetic contamination (unexpected outcrossing) occurs in an animal room containing two strains with the same coat color, the genetically contaminated animals cannot be differentiated visually. To compound the problem, in many cases these animals show a high level of propagation because they are healthier than average and their reproductive capacity is improved by the hybrid vigor.

Biomedical researchers must assume that the animals they purchase and use are genetically pure: should the animals be contaminated, the results of experiments using them would be thrown into doubt; such studies would not be comparable with other experiments conducted under the same conditions using the same strains. Not only are the experimental results worthless in such cases, but the time, cost and effort spent on the experiments are wasted, and the researcher's trustworthiness as a scientist will be damaged. Researchers must use animals whose genetic quality is checked scientifically and assured objectively, to obtain reliable results in animal experiments.

Genetic monitoring is a system which backs up genetic control and objectively assures genetic quality by scientific checking on the course of maintenance, production and supply of laboratory animals.

In the rearing of laboratory animals which serve as biological measuring tools in experiments in such fields as biology, medicine and pharmacology, all factors in the environment are strictly monitored by means of microbiological and nutritional control and regulation of rearing conditions. Not only drugs, food and feed but also almost all other industrial products used are supplied only after passing through a strict course of quality control. By comparison, the monitoring of genetic factors has lagged behind. Even where laboratory animal facilities are excellent, major mistakes may occur when there is inadequate genetic control or monitoring.

Genetic monitoring must fulfil four conditions simultaneously: it must be (1) exact, (2) easy, (3) efficient and (4) economical. Therefore, the genetic characters subject to genetic monitoring and the methods used to check them must be selected to meet these conditions.

The selection of genetic characters is very important since it has a direct effect on the reliability of genetic monitoring. Characters which cause major changes in the phenotype due to differences in environmental factors such as nutritional conditions, rearing temperature and microbial infections cannot be used. In selecting the methods used to check the genetic characters, it is also essential to choose those which give exact results. Methods which give results with low reproducibility must be avoided.

It is also important that it be easy to distinguish the genetic character. Genetic contamination must be easily identifiable, and monitoring results must be explained scientifically and understood easily even by laymen. The checking methods must not require high-level techniques or complex procedures. Ideally, ordinary animal caretakers in animal houses should be able to perform these procedures.

Genetic monitoring must also be economical to carry out. In addition, it is desirable to be able to distinguish the genetic characters with no sacrifice or injury to the animals; to avoid methods which use expensive chemicals, especially those which are difficult to dispose of, require expensive equipment, or need technicians with special licenses; and to be able to detect several genetic characters using one technique.

This volume recommends to biomedical researchers a system and procedures for carrying out genetic monitoring of their laboratory animals. It lists principles of good colony management; describes the genetic characters that should be monitored in laboratory animals, emphasizing coat color, biochemical markers, and immunogenetic markers; and describes techniques for genetic observation and analysis.

I. GENETIC MANAGEMENT AND EMBRYO BANKING

Colony Management

Good colony management incorporates a number of practices that contribute to the prevention of accidental genetic contamination. In general, any measures that can be taken to prevent unwanted animals from entering a breeding cage are paramount. Specific practices that can be observed in this regard include the following:

1. Animal rooms should be clean and uncluttered. Cages should be kept in good repair to prevent animals from escaping or strays from entering.

2. Animals of separate strains that have the same coat color and that are otherwise indistinguishable should be housed in separate locations in so far as possible.

3. Cage tags used to identify animals should be color-coded by strain to prevent accidental mix-ups.

4. Cages should be cleaned one at a time to prevent the accidental exchange of animals.

5. Animal caretakers should be given orientation training that includes basic genetics as well as animal handling techniques. Such training will help instill an appreciation of the consequences of incorrect breeding.

6. Animal caretakers should be trained and given incentives to search for and report any deviations in numbers or phenotypes of animals. Litters having more animals at the time of weaning than recorded at birth are suspect and should be discarded.

7. Animal caretakers should be given instructions to destroy or isolate any animal found outside a cage. A number of traps can be set around an animal room to catch strays.

8. Only highly qualified personnel should be given the responsibility for handling animals that make up a foundation colony. All procedures in that colony should be closely supervised.

9. Any precipitous change in breeding performance should be considered suspect, and breeders should be thoroughly typed for genetic markers.

Maintenance and Expansion of Inbred Strains

An inbred strain is one derived from 20 or more consecutive generations of brother × sister mating (or in exceptional cases younger parent × offspring mating), with all individuals of the colony being derived from a single breeding pair in the 20th or a subsequent generation. Such strains have a number of useful properties not found in normal outbred stocks. These properties are discussed in detail by Festing (1979), but the more important ones in relation to genetic quality control may be summarized very briefly as follows:

1. Members of an inbred strain are *isogenic*, i.e., they are genetically identical to each other. This is of immense importance as it forms the basis of genetic quality control programs based on single-gene markers. Thus, a single individual can be typed, for example, at the *Hbb* locus, and this will imply that all members of the same inbred strain have the same genotype at that locus. Any animal subsequently found not to have the stated genotype would immediately be suspected of being the result of genetic contamination or of coming from a different strain.

2. All individuals of an inbred strain should be *homozygous* at more than 99% of all genetic loci. Such animals should carry virtually no hidden recessive genes, and so will breed true if mated to an individual of the same inbred strain. This property is again of importance in genetic quality control programs, as any individual found to be heterozygous at any locus would immediately be suspected of being the result of genetic contamination.

3. An inbred strain should stay *genetically constant* for long periods of time. Genetic change can only arise as a result of "residual heterozygosity" as inbreeding progresses and as a result of new mutations accumulating and becoming established in the strain. Change as a result of these two influences, while important

5

in some contexts, is relatively slow, so that it is valid to consider that inbred strains remain constant for long periods of time. This again is important in genetic quality control programs as it means that data collected on the genotype of a strain at a particular locus should remain valid for many years.

4.   Each inbred strain is a *unique combination* of some 30,000 genetic loci and will therefore differ from every other inbred strain. This is important because each strain is likely to have characteristics which are of interest to some group of research workers. From the point of view of genetic quality control, this property is important because each strain can be uniquely typed either by studying an array of polymorphic loci such as those controlling electrophoretic variants of enzymes and proteins or by studying morphological or other characteristics of the strain that can be used to distinguish it from all other strains.

5.   Other properties of inbred strains which make them useful are their international distribution and their sensitivity to environmental influences, though the latter property is not well defined.

Principles of strain maintenance

The main aim of a breeding colony is to produce experimental animals of a specified type that remain true to their specification for as long as possible. Therefore, the three factors which can lead to genetic change—namely, genetic contamination, mutation, and drift due to residual heterozygosity—should be controlled as closely as possible.

Unfortunately, little can be done to control drift due to residual heterozygosity. However, although such drift will be important prior to and to a lesser extent even after 20 generations of brother × sister mating, it is unlikely to be of much importance in strains maintained, for example, for 40 or more generations of such mating (Bailey, 1978). Most of the standard inbred strains were produced in the 1930s and have probably been inbred for well over 100 generations of brother × sister mating.

The accumulation of mutation is difficult to avoid, though some success may be achieved by choosing only animals of normal appearance to maintain the colony and by selection for high breeding performance. This is justified on the grounds that most mutation that are not visible will have a deleterious effect on "fitness," which is best measured through breeding performance. Selection for good breeding performance in an inbred strain would not be expected to increase such performance, but it might prevent it from decreasing due to the fixation of mildly deleterious mutations; at the same time this will help to prevent these mutants from becoming fixed in the colony. However, selection for good breeding performance should be conducted with care, as genetic contamination may result in some individuals expressing "hybrid vigor" or "heterosis," which results in extremely good breeding performance. Such contamination could spread rapidly through the colony if selection for good breeding performance is applied uncritically. These problems are discussed in more detail below.

Genetic change due to genetic contamination is likely to be several orders of magnitude more serious than genetic changes due to the other two causes, but fortunately it should be relatively easy to detect and of course it should be preventable. In fact, *prevention of genetic contamination should be one of the main aims of those in charge of a breeding colony.* This can be achieved in the following ways:

1.   Physical separation of each breeding colony should be maintained as far as practicable. Particular attention should be paid to physical separation of the foundation stock colony (see below). In some animal facilities, for example, the foundation stock of each inbred strain is maintained in a different isolator so that it is physically impossible for two strains to become muddled.

2.   Where complete physical separation of each colony is impossible, the colonies should be segregated by coat color insofar as possible. Thus, if it is essential to maintain, for example, 3 colonies in an animal room, these should be chosen so that they each have a different coat color.

3.   If many foundation stock colonies must be maintained in the same animal room, then physical separation by keeping each strain in a different area and alternating coat colors should be used to keep the colonies separate. In such circumstances each colony should have a distinctive color-coded cage label, and there should be an absolute rule that any animal that escapes will be destroyed.

4.   A breeding system should be chosen that minimizes the movement of animals. In particular, wherever possible mice should be bred as monogamous pairs or trios (depending on the type of colony) that remain together for their entire breeding lives, rather

than being kept as harems in which pregnant animals are removed from the group prior to parturition. The latter system, involving considerable movement of animals, is much more likely to result in mistakes being made.

Subdivisions of breeding colonies of inbred strains

If animals of an inbred strain are to be produced in large numbers (e.g., more than about 50 mice per week), it is usual to subdivide the breeding colony into one or more of the following:

1. *Foundation stock colony (FS).* This colony represents the "seed corn" of the strain. The size of the FS colony is usually quite small, ranging from about 10 to 50 breeding pairs, and all animals are bred as permanently mated monogamous pairs. The young produced by this colony are used first to maintain the colony, and second as breeding stock for one of the other types of colony listed below. As the FS colony is the only one that contributes to the long-term survival of the strain, it is clear that some sort of routine genetic quality control procedure should be used within this colony, especially if it is not physically separated from all other strains.

2. *Multiplication or expansion colonies (MC).* The FS colony provides breeding stock for one or more types of multiplication or expansion colony, depending on individual circumstances. The aim of these colonies is to produce breeding stock of the specified type for the final production of experimental animals, but these colonies do not normally contribute to the long-term survival of the strain. Exceptionally, if there were an accident leading to the loss of the FS colony, then the strain could be perpetuated via one of the multiplication colonies. The following types of MC are used in various institutes:

a. "Pedigree expansion stocks" (PES). Such colonies, which are maintained by brother × sister mating and in which full details of the pedigree and performance of each pair is maintained, are often favored when very large numbers of experimental animals are to be produced. Such colonies are also commonly maintained by institutes with a special interest in mouse mutants, as the origin of the mutation can easily be traced.

b. "Traffic light" expansion colonies. The traffic-light system was proposed by Lane-Petter as a simple means of keeping track of the number of generations of expansion, which he suggested should be limited to 3 generations. Foundation stock go in cages with white labels, and the three generations of expansion (which may be by random mating) go in boxes with green, yellow, and finally red labels. The output of the boxes with red labels are used solely for experimental purposes. None of the offspring of the traffic light scheme would contribute to the long-term survival of the strain under normal circumstances.

c. "Production colonies." Experimental animals may be produced from what some authorities call a production colony, whose aim is only to produce experimental animals.

Although this division of the breeding colony might appear to be rather complex, it should be emphasized that the main division is between the FS colony, which contributes to the long-term survival of the strain, on the one hand, and the multiplication/production colonies, which only result in short-term expansion and the production of experimental animals, on the other. As the various multiplication colonies do not contribute to the long-term survival of the strain, continued brother × sister mating in such colonies is optional. However, it is vital in the FS colony in order to maintain homozygosity. Brother × sister mating in the FS colony is also important for genetic quality control. Random mating of such a colony would mean that a low level of genetic contamination of, say, one animal among 30–100 animals would be very difficult to detect. With brother × sister mating such genetic contamination would be revealed much more easily as it would be confined to a single branch of the pedigree chart (see below).

FS (foundation stock) colony maintenance

The FS colony should normally consist of about 10 to 50 monogamously mated pairs of animals. Each cage should have a color-coded label. Cages should normally be numbered consecutively according to the date of mating. As the pair is always made up by mating brother × sister, the individuals do not need to be assigned separate pedigree numbers. Individual identification by ear punching or tattooing is optional, but is not usually necessary when monogamous pair mating is used. Each cage should have a record card showing the following information: pair number, date of birth, date of mating, pair number of parents, and details of litters born and number of young weaned (see Fig. 1).

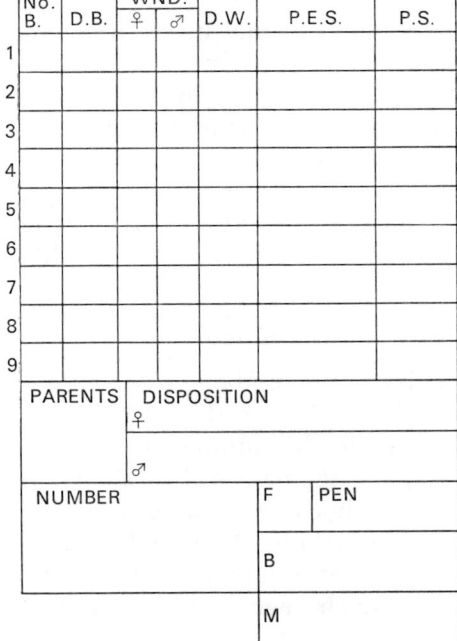

BOX No		WEIGHT	BORN	FROM	F	REASON FOR DISPOSAL (OF ♀)	TICK
	♂					DEATH OR DISEASE OF ♂ OR ♀	
MADE UP ON	♀					POOR BREEDING PERFORMANCE	
♀ CULLED ON	♀					END OF BREEDING LIFE	
						OTHER	

No	AT BIRTH			AT WEANING				REMARKS
	DAY BORN	NO BORN	CUM TOTAL	♂♂	♀♀	CUM TOTAL	AV WEIGHT	
1								
2								
3								
4								
5								
6								
7								
8								
9								
10								
11								
12								
13								

No. B.	D.B.	WND. ♀	WND. ♂	D.W.	P.E.S.	P.S.
1						
2						
3						
4						
5						
6						
7						
8						
9						

PARENTS	DISPOSITION	
	♀	
	♂	
NUMBER	F	PEN
	B	
	M	

FIG. 1.   Examples of cage labels suitable for inbred strains of mice.

Actual designs of record cards vary by institute. Two examples are shown in Fig. 1. The record card may be attached directly to the cage, or it may be kept separately. In the case of computerized records, the cage record would be kept in the computer and would be accessed via the pair number.

In addition to the cage record, each colony should also have a pedigree chart showing the relationship between pairs. An example of such a chart is shown in Fig. 2. The aim of this chart is to ensure that all animals trace back to a common ancestral pair within 5–7 generations. This prevents wide divergence within the colony arising as a result of mutational drift. The pedigree chart

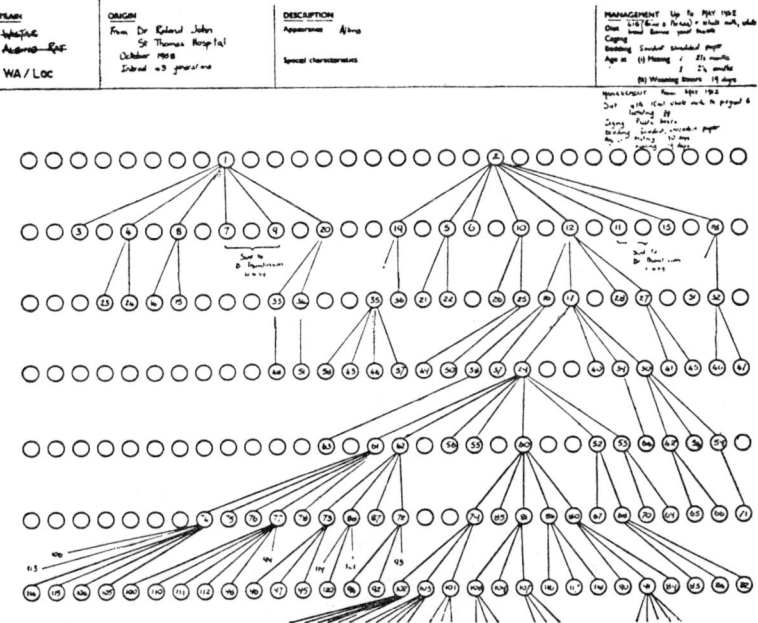

FIG. 2. Example of a pedigree chart using a preprinted form. Note that each circle represents a monogamous breeding pair. The chart needs to be modified for those species which are not bred as monogamous pairs. Pair numbers are assigned consecutively. Details of the origin, husbandry, dispatch, cross-fostering, etc., of the strain may be kept conveniently on such a chart. This chart refers to May 1962. Husbandry changes since then have been recorded on subsequent charts (Festing, 1979).

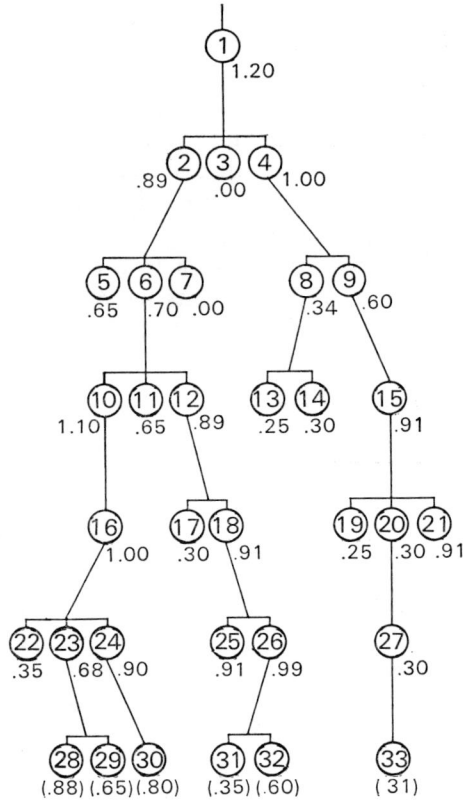

FIG. 3. Example of a hypothetical pedigree chart to illus-
trate selection of future breeding stock in the FS
colony. Pair numbers are given within the circles,
and productivity (young/female/wk) beside each
circle. The numbers in parentheses in generation
7 are provisional as these pairs are still breeding.
Note that all pairs trace back to a common ancestor
6 generations back (pair No. 1), and there are two
major sublines arising from pairs 2 and 4. It is
recommended that only one of these sublines be
propagated. Choice of subline should be based on
the mean breeding performance of the whole subline rather than the performance of individuals in the
seventh generation. The mean of the subline derived
from pair 2 is:

$$\frac{0.89+0.65+ \ldots +0.60}{20}=0.71$$

may also be used in the selection of future breeding stock. In order to avoid fixing deleterious mutant genes and a decline in breeding performance, sublines should be selected for good breeding performance. This may be carried out most efficiently by considering the average breeding performance of the whole subline (past as well as present). An example is given in Fig. 3. Briefly, the method is to calculate the average breeding performance of all branches of the pedigree that trace back to the common ancestral pair and choose breeding stock from that branch with the best breeding performance. Obviously all animals which are abnormal in appearance or behavior should also be discarded from the breeding colony. It must be remembered that if genetic contamination has occurred, the resulting hybrid young will probably exhibit vigor for breeding performance. Therefore any animal exhibiting exceptionally good breeding performance should be treated with suspicion. In fact, monitoring of breeding performance can be used as a preliminary screening method for detecting genetic contamination. This is discussed in more detail in the chapter on the use of quantitative characters for genetic quality control.

The pedigree chart should also be used in the selection of animals for routine genetic quality control. Ideally, such sampling should be arranged so that all mice that subsequently become the common ancestors of the whole colony have been subjected to some genetic quality control screening procedure. Alternatively, genetic quality may be assessed by sampling offspring from breeding pairs which are as far apart on the pedigree chart as possible. In any case, the pedigree chart is a convenient place on which to record the sampling and results of any routine screening that has been carried out.

The mean of the subline derived from pair 4 is:

$$\frac{1.00+0.34+0.60+ \ldots +0.31}{11}=0.50$$

Thus, future breeding stock should be taken from the subline derived from pair 2. Note that very high levels of productivity should be treated with suspicion as they may be due to hybrid vigor following genetic contamination. Special priority should be given to sublines that have been screened for genetic authenticity (Festing, 1979).

Multiplication and production colony maintenance

An FS colony of only a few breeding pairs may form the peak of a pyramid that results in the production of many thousands of mice following several generations of multiplication. Thus, although the maintenance of FS colonies is relatively expensive, this cost represents only a small fraction of the final cost of an experimental animal. Fortunately, the maintenance of the various multiplication colonies requires much less detailed records, and there is no particular merit in continued brother × sister mating in such colonies, though (as mentioned earlier) a pedigree expansion stock may be worth maintaining under some circumstances.

In general, great care should be taken to prevent genetic contamination of expansion and production colonies, but the only records that need to be kept are those that are essential for efficient colony management. Thus, a system should be devised to ensure that nonproductive stock are culled and that breeding animals are not kept beyond their normal economic breeding life spans. Records of the total output of the colony on a weekly or monthly basis should also be kept as a check on the overall productivity of the strain. Such records are essential if the colony is to be managed efficiently. If individual cage records are kept, breeding performance can be monitored with the aim of evaluating hybrid vigor.

The role of animal technicians and caretakers

Although some scientists look after their own inbred strains personally, most colonies are under the day-to-day care of an animal technician or caretaker. It is obvious that if more than one strain is maintained in any animal room the chance of its becoming genetically contaminated and the chance of such contamination being detected quickly will depend to a very large extent on the personal qualities and training of the technician. Foundation stocks should only be looked after by technicians who understand and care about their potential value in research. Such technicians should be well trained in general aspects of laboratory animal husbandry, but they should also be given specific training in the maintenance of inbred strains. In particular, they must be shown how to maintain the records correctly, and they should be trained to recognize any animal that is abnormal with respect to growth, reproduction, behavior, or morphology. A technician who succeeds in detecting an instance of genetic contamination through such observation should be rewarded with praise rather than being blamed for causing the contamination in the first place. In this way caretakers will be encouraged to work to the best of their ability in the future.

Embryo Freezing and Banking

Frozen embryo storage provides a secure and economical backup for stock used in research. Conventionally bred stocks are subject to loss through disease or environmental accident, and to change through genetic drift or genetic contamination. Thus, although frozen embryo storage cannot prevent incorrect matings, it does offer a means of rescue from otherwise irreparable genetic contamination.

Embryos can be successfully frozen by several different protocols that utilize a range of equipment from the relatively simple and inexpensive to more complex, automated, and expensive apparatus. Details of such protocols can be found in the references listed in this manual. The protocol used and described previously by Mobraaten (1981) will be given here. The procedure of collecting, freezing, and thawing embryos is basically that originally described by Wittingham, Leibo, and Mazur (1972), with the exception that an automated freezing rate controller is used.

Collection

Female mice from which embryos will be taken are given injections of the gonadotropins PMS (pregnant mare's serum) and HCG (human chorionic gonadotropin) to obtain consistently large numbers of embryos at a predetermined time. The optimum dose of gonadotropins varies with the strain, but generally 2.0 IU of PMS and 2.5 IU of HCG given to females between 21 and 28 days of age, according to the schedule in Table 1, give good results.

Eight-cell embryos are aseptically collected on day 3 (appearance of the vaginal plug is taken as day 1) from donors mated

16

TABLE 1. Schedule of gonadotropin injections

Day	Procedure	Time	Hours after HCG injection
0	PMS injection	1500	−48
1	—	—	—
2	HCG injection	1500	0
	Mating with male	1530	0.5
3	Plug check	0800	17
4	—	—	—
5	Embryo collection	0800	65

after gonadotropin-induced ovulation. On the day of embryo collection, an oviduct and part of the anterior horn of the uterus are removed from a female killed by cervical dislocation and put in a drop of phosphate-buffered saline (PBS) at pH 7.2. Then, by introducing a 30-gauge needle attached to a 5-ml syringe filled with PBS into the uterine end of the oviduct, the embryos are flushed from the oviduct into an embryological watchglass or depression slide. A clutch of embryos from each female is kept separate during manipulation in order to maintain pedigree identity.

Preservation

After flushing, the embryos are transferred by means of a specially drawn out pipette to 2-ml plastic cryotubes containing 0.1 ml of PBS. The ampules are then placed in an ice bath (0°C) and held there for 15 minutes to allow equilibration of temperature. An additional 0.1 ml of PBS with 2 M dimethyl sulfoxide (DMSO) is then added to bring the final concentration of DMSO to 1 M. After another 15 minutes' equilibration at 0°C, the ampules are transferred to a salt and ice bath maintained at −6°C. Two minutes after immersion in the ice bath, the contents of the ampules are seeded (induction of ice formation) with an ice crystal from the tip of a Pasteur pipette. The ampules are then capped and transferred to a Linde CRFC-2 freezing chamber that has been precooled to −6°C. The temperature is lowered at the rate of approximately 0.5°C per minute by means of a Linde CRC-1 controller that regulates the amount of liquid nitrogen injected into the chamber. Temperature changes in the freezing chamber are monitored by the use of a thermocouple inserted in a control ampule and connected to a temperature recorder. When the temperature of the freezing chamber has been lowered to $-80°C$, the ampules are removed from the freezing chamber and immediately put into a liquid nitrogen refrigerator.

Thawing

The thawing of embryos is accomplished by removing the ampules from the liquid nitrogen refrigerator and placing them in the freezing chamber, which has been precooled to $-100°C$. Cessation of the liquid nitrogen and activation of the heating element in the chamber results in a warming rate of approximately $8°C$ per minute. The temperature is monitored until it reaches $0°C$, whereupon the samples are allowed to liquify. 0.8 ml of PBS is added to dilute the DMSO, because it has been shown that DMSO is toxic to embryos at temperatures above $0°C$, and therefore dilution is necessary before the medium containing the embryos warms up to room temperature (Whittingham and Wales, 1969). The contents of the ampules are withdrawn by means of a micropipette and transferred to embryological watchglasses or depression slides where the embryos can be observed. The embryos are then transferred into 12×75-mm tubes containing about 1 ml of Whitten's medium (Whitten, 1971) that has been gassed for 20 seconds with a gas mixture of 5% carbon dioxide, 5% oxygen, and 90% nitrogen. The tubes are then put in a $37°C$ incubator for about 15 minutes to allow for equilibration between the medium and the gas, and gassed a second time just prior to the introduction of embryos. After the embryos are introduced, the tubes are gassed one more time and sealed. Embryos are cultured in the test tubes overnight, and their viability is ascertained the following morning.

Transfer to pseudopregnant mothers

To recover live-born young, thawed and cultured embryos are transferred into pseudopregnant female mice; F_1 hybrid mice generally give the best results as recipients. Pseudopregnant mice are produced by selecting females in proestrous and mating them to vasectomized males that have been proven sterile. Embryos, in the blastocyst stage after culturing, are transferred to these females on the third day of pseudopregnancy (vaginal plugs are observed on day 1).

The transfer is carried out by injecting the blastocysts into the uterus. The uterus is exposed by making a small incision through the skin on the midline of the back at the level of the kidneys of the anesthetized, pseudopregnant recipient. The incised opening is then manipulated to a position just over one or the other of the ovaries (just below the kidney), where another small incision is made in the peritoneum. The uterus can then be found, withdrawn sufficiently, and held in place or manipulated by means of sterile thread. The injection of embryos is carried out by means of a finely drawn-out pyrex pipette whose tip has been briefly fire-polished. The pipette is inserted through an incision made in the uterine wall about 5 to 10 mm from the oviduct with a 25-gauge hypodermic needle. The embryos, in culture medium, are drawn just into the tip of the pipette and injected into the uterus with a minimum amount of medium. The uterus is placed back into the body cavity, and two wound clamps are affixed to the skin to close the wound.

A minimum of 500 embryos should be collected from each strain to be stored to insure an adequate number for subsequent recovery. Embryos that are collected from any given strain should be placed in two separate liquid nitrogen refrigerators for safety. It is important to assure that an adequate level of liquid nitrogen is maintained in the refrigerators at all times. Many refrigerators can be fitted with liquid level alarms that sound when the level of liquid nitrogen falls below a pre-set level. A continual supply and an adequate reserve should be maintained to allow a wide margin of safety in the event of a delay in the regular delivery of liquid nitrogen.

Since the chief purpose of a frozen-embryo bank is to safeguard materials over a period of time during which there will undoubtedly be changes in personnel and methods, it is imperative that complete and accurate records be maintained. These are essential to the identification and recovery of stored materials by future guardians or users.

Record-keeping for embryo banking

It is recommended that the following records be maintained:

(1) Conditions of freeze run. A log should be kept of the physical conditions, including the protocol used, for freezing the embryos along with the results of viability tests carried out on a sample of embryos recovered from each run. Such records will be needed not only to document the fact that the conditions of the freezing were correct but also to indicate what the correct protocol will be for thawing the embryos. The method to be used for thawing embryos depends critically on the method used for freezing.

(2) Stock description. For each strain maintained in the repository, the origin, phenotype, breeding performance, response to the procedures of embryo freezing, and other information that might be of use to future users should be kept on file.

(3) Genetic identification and storage location of embryos. Strain, genotype, pedigree numbers of parents, and generation of inbreeding can be recorded. For any embryo frozen from a pedigreed stock, sufficient information should be recorded to enable its ancestry to be traced to the colony from which it was derived. The code number on the vial in which the embryos are contained and its precise location in the liquid nitrogen refrigerator should also be recorded.

References

Bailey, D. W. (1978). Sources of subline divergence and their relative importance for sublines of six inbred strains of mice. *In* H. C. Morse III (ed.) Origin of Inbred Mice. Academic Press, New York, pp. 197–216.

Festing, M. F. W. (1979). Inbred Strains in Biomedical Research. Macmillan Press, Ltd., London.

Mobraaten, L. E. (1981). The Jackson Laboratory genetics stocks resource repository. *In* G. Zeilmaker (ed.) Proceeding of the Workshop on Embryo Storage and Banking in Laboratory Animals. Gustav Fischer Verlag, Stuttgart, pp. 165–177.

Whitten, W. K. (1971). Nutrient requirements for the culture of preimplantation embryos *in vitro*. *Advances in Bioscience*, **6**: 129–141.

Whittingham, D. G. and R. G. Wales (1969). Storage of two-cell mouse embryos *in vitro*. *Austr. J. Biol. Sci.*, **22**: 1065–1068.

Whittingham, D. G., S. P. Leibo, and P. Mazur (1972). Survival of mouse embryos frozen to −196 and −296°C. *Science*, **178**: 411–414.

II. QUALITATIVE CHARACTERS IN GENETIC MONITORING

Qualitative genetic characters are generally those in accordance with simple Mendelian law: there is good correspondence between the phenotype and the genotype, and the site of locus of the gene and correlations among loci on the chromosomes are often clear.

However, it is impossible to perform genetic monitoring with only one genetic character, and various techniques are used since it is essential to examine a rather large number of loci simultaneously. Figure 4 shows the qualitative genetic characters used in genetic monitoring for each chromosome (Hoffman *et al.*, 1980). If characters which are clear genetically are selected, clear results will be obtained for each animal and genetic contamination can be easily checked.

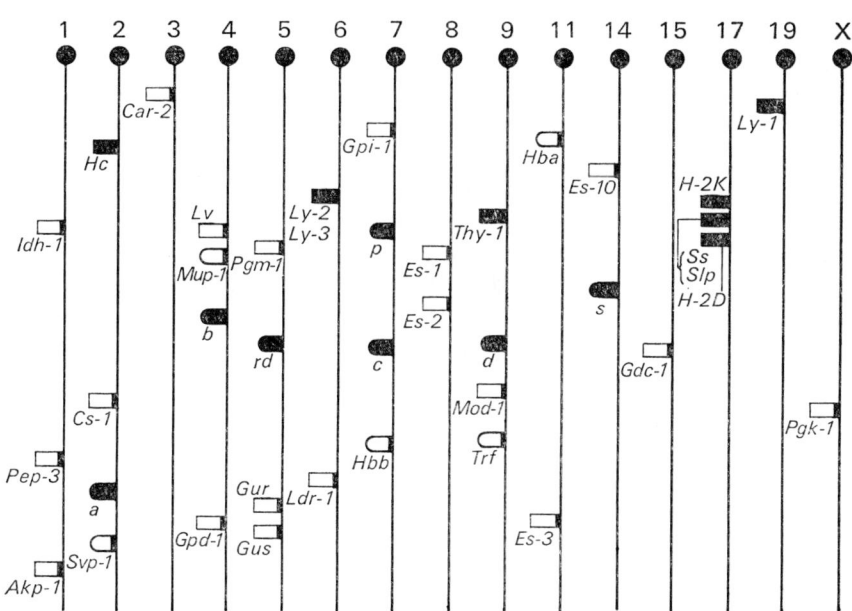

FIG. 4. Linkage map of the genetic markers used for genetic monitoring

▇ Morphological   ▇ Immunological
▢ Enzyme   ◖ Protein

Choice of Characters

Coat color

Of the large number of genetic variations in the coat colors of mice (Silvers, 1979), only a limited number are used in genetic monitoring. The rare variants that are found in peculiar strains are known as strain-specific coat color and are used directly in genetic monitoring.

Coat color markers routinely used for genetic monitoring are shown in Table 2. These are easily classified in an animal house without special equipment. However, fundamental knowledge of animal genetics is required to promote genetic monitoring procedures.

The strain distribution of coat colors is given in Table 3. There is a strong preference for albino strains as laboratory mice, and this causes serious problems in terms of the masking of genetic contamination, because of the large number of strains which have white coats.

Biochemical markers

The number of genetic variations of enzyme (isozyme) and protein in the mouse is large (Roderick *et al.*, 1981). However, only a limited number of these are used as biochemical markers in genetic monitoring because the rare variations among strains are not efficient for use in routine genetic monitoring.

Biochemical markers that are frequently used for genetic monitoring are shown in Table 4. All of these markers can be detected by electrophoresis and histochemical staining. The strain distribution of variations in several loci are given in Table 5. In this

TABLE 2. Coat colors used for genetic monitoring

Locus symbol and name		Alleles	Chromosome No.
c	Colored-albino	C, c^{ch}, c^h, c, etc.	7
a	Agouti-nonagouti	A^y, A^w, A a^t, a, etc.	2
b	Black-brown	B, b, etc.	4
d	Dilution	D, d, etc.	9
p	Pink-eyed dilution	P, p	7

TABLE 3. Genetic profiles of coat color markers

Strain	Color	Chromosome	2	4	7	9	–
		Locus	a	b	c	d	$*$
A/He	White		a	b	c	D	
AKR	White		a	B	c	D	
BALB/c	White		A	b	c	D	
CBA/J	Black-agouti		A	B	C	D	
C3H/He	Black-agouti		A	B	C	D	
C57BL/6	Black		a	B	C	D	
C57BL/10	Black		a	B	C	D	
C57BR/cd	Brown		a	b	C	D	
C57L	Leaden		a	b	C	D	$ln(1)$
C58	Black		a	B	C	D	
DBA/1	Dilute brown		a	b	C	d	
DBA/2	Dilute brown		a	b	C	d	
DDK	White		A	B	c	D	
KK	White		a	B	c	D	
NC	Brown-agouti		A	b	C	D	
NZB	Black		a	B	C	D	
NZW	White		A	b	c	D	
RIII	White		A	B	c	D	
SJL	White		A	B	c	D	$p(7)$
129	Light chinchilla		A^w	B	c^{ch}	D	$p(7)$

$*$: Additional locus (chromosome)
(a=non agouti; b=brown; B=black; c=albino d;=dilute; ln=leaden; p=pink-eyed dilution; A^w=white-bellied agouti; c^{ch}=chinchilla)

table, a letter a shows that the genotype is homozygous for a/a, b shows homozygous for b/b, and so on (i.e., the genotype of $Idh-1$ locus in the strain AKR is b/b).

The genotype pattern of many loci is called the genetic profile of the strain. This profile is useful for the identification of strains when it is compared with the results of genetic monitoring.

Immunogenetic markers

A large number of immunogenetic markers can be used for genetic monitoring (Tables 6 and 7). However, only a certain number of these markers prove feasible as routine measures in genetic quality assurance. In the following, an introduction to these markers will be given. The advantages and disadvantages of

TABLE 4.   Biochemical markers frequently used for genetic monitoring

Locus symbol and name		Allels	Chromosome No.
Akp-1	Alkaline phosphatase-1	a, b	1
Car-2	Carbonic anhydrase-2	a, b	3
Ce-2	Kidney catalase-2	a, b	17
Es-1	Esterase-1	a, b	8
Es-2	Esterase-2	a, b, c	8
Es-3	Esterase-3	a, b, c	11
Es-10	Esterase-10	a, b, c	14
Gpd-1	Glucose-6-phosphate dehydrogenase-1	a, b, c	4
Gpi-1	Glucose phosphate isomerase-1	a, b	7
Hbb	Hemoglobin beta-chain	d, p, s	7
Idh-1	Isocitrate dehydrogenase-1	a, b	1
Ldr-1	Lactate dehydrogenase regulator	a, b	6
Mod-1	Malic enzyme supernatant	a, b	9
Mup-1	Major urinary protein	a, b	4
Pep-3	Peptidase-3	a, b, c	1
Pgm-1	Phosphoglucomutase-1	a, b	5
Trf	Transferrin	a, b	9

these markers for this purpose will be discussed in detail in the section on Laboratory Procedures.

Immunogenetic markers are classified according to their nature and function into cell membrane-associated alloantigen loci (erythrocyte alloantigen loci, differentiation alloantigen loci, and histocompatibility loci) and allotypes that are genetically determined antigenic determinants of soluble proteins in the serum varying within the same species (immunoglobulin allotypes, serum protein allotypes).

Cell membrane-associated alloantigen

Erythrocyte alloantigen loci: The antigens coded for by these loci are primarily expressed on erythrocytes and are serologically

TABLE 5. Genetic profiles in biochemical markers

Chromosome	1	1		3		4	5	6	7	7	8	8	9	9	11	14	17
Locus	Idh	Pep	Akp	Car	Mup	Gpd	Pgm	Ldr	Gpi	Hbb	Es	Es	Mod	Trf	Es	Es	Ce
	-1	-3	-1	-2	-1	-1	-1	-1	-1		-1	-2	-1		-3	-10	-2
A	a	b	b	b	a	b	a	a	a	d	b	b	*	b	c	a	a
AKR	b	b	b	a	a	b	a	a	a	d	b	b	*	b	c	b	b
BALB/c	a	a	b	a	a	b	a	a	a	d	b	b	a	b	a	a	a
CBA	b	b	a	*	a	b	*	*	b	d	b	b	b	a	c	b	b
CL/Fr	a	b	b	a	a	b	b	a	a	d	b	b	a	b	c	·	·
C3H/He	a	b	b	b	a	b	b	a	b	d	a	b	b	b	c	b	b
C57BL/6	a	a	a	a	b	a	a	a	b	s	a	b	a	b	a	a	a
C57BL/10	a	a	a	a	b	a	a	a	b	s	a	b	b	b	a	a	a
C57L	b	a	a	b	b	a	b	a	a	s	b	b	a	b	a	a	a
DBA/1	b	b	a	a	a	a	b	a	a	d	a	a	a	b	c	·	·
DBA/2	b	a	a	a	a	b	b	a	b	d	b	a	a	b	c	b	a
KK	a	b	b	a	b	a	a	a	a	s	b	a	a	b	c	·	a
NC	b	b	a	a	b	b	b	a	a	s	b	b	b	b	c	·	·
NZB	a	c	*	a	*	b	b	a	a	*	*	b	a	b	*	b	b
NZW	b	b	b	a	a	b	b	a	a	d	b	b	b	b	c	·	·
P/J	b	c	b	a	b	a	b	a	a	d	b	b	a	b	a	·	·
RF/J	a	b	b	a	a	a	a	a	a	d	b	b	b	b	b	b	b
RIII	*	b	·	b	*	b	b	a	a	*	b	b	b	*	*	·	·
SJL/J	b	b	b	b	a	a	b	a	a	s	b	b	a	b	c	b	a
129	a	b	b	a	a	a	a	a	a	d	b	b	a	b	c	b	b

*: Not all substrains have the same allele.

Table 6.   Immunogenetic markers in inbred strains of mice

Locus symbol and name		Allele	Chromosome No.
Ea-1	Erythrocyte antigen-1	a, b, o	8
Ea-2	Erythrocyte antigen-2	a, b	—
Ea-3	Erythrocyte antigen-3	a, b	—
Ea-4	Erythrocyte antigen-4	a, b	—
Ea-5	Erythrocyte antigen-5	a, b	—
Ea-6	Erythrocyte antigen-6	a, b	2
Ea-7	Erythrocyte antigen-7	a, b	—
Ea-8	Erythrocyte antigen-8	a, b	—
Hc	Hemolytic complement	1, 0	2
H-1 to H-39		See Table 7	
H-X	Histocompatibility-X	See Table 7	
H-Y	Histocompatibility-Y	See Table 7	
Igh-C	Igh constant region	6 loci	12
Igk-V	Igk variable region	3 loci	6
Igl-C	Igl constant region (Igl-1)	a, b	—
Ly-4	Lymphocyte antigen-4	a, b	2
Ly-5	Lymphocyte antigen-5	a, b	1
Ly-6	Lymphocyte antigen-6	a, b	9
Ly-7	Lymphocyte antigen-7	a, b	—
Ly-8	Lymphocyte antigen-8	a, b	—
Lyb-2	B-lymphocyte antigen-2	a, b, c	4
Lyb-4	B-lymphocyte antigen-4	a, b	4
Lyb-5	B-lymphocyte antigen-5	a, b	—
Lyb-6	B-lymphocyte antigen-6	a, b	—
Lyt-1	T-lymphocyte antigen-1	a, b	19
Lyt-2	T-lymphocyte antigen-2	a, b	6
Lyt-3	T-lymphocyte antigen-3	a, b	6
Qa-1	Qa lymphocyte antigen-1	a, b	17
Qa-2	Qa lymphocyte antigen-2	a, b	17
Qa-3	Qa lymphocyte antigen-3	a, b	17
Sas-1	Serum antigenic substance	a, o	1
Thy-1	Thymus cell antigen-1	a, b	9
Tla	Thymus leukemia antigen	a, b c, d	17

detectable by the standard blood typing techniques (Snell and Cherry, 1972). Table 8 gives the genetic profiles of several erythrocyte alloantigens. *Ea-5* and *Ea-6* readily distinguish between the Heston and Strong sublines of C3H. Unfortunately there is no detailed strain distribution pattern elaborated, and only the loci *Ea-1* (which is not polymorphic in inbred strains) and *Ea-6* have been assigned to chromosomes 8 and 2, respectively. For the other antigens, no linkage data have yet been obtained.

TABLE 7. Histocompatibility loci of inbred mouse strains

Locus symbol and name		Allele	Chromosome No.
H-1	Histocompatibility-1	a, b, c, ?	7
H-2	Histocompatibility-2	see Table 10	17
H-3	Histocompatibility-3	a, b, c, d, e, f	2
H-4	Histocompatibility-4	a, b	7
H-7	Histocompatibility-7	a, b	9
H-8	Histocompatibility-8	a, b, c	—
H-9	Histocompatibility-9	a, b	—
H-10	Histocompatibility-10	a, b	—
H-11	Histocompatibility-11	a, b ?	—
H-12	Histocompatibility-12	a, b	—
H-13	Histocompatibility-13	a, b, c	2
H-15	Histocompatibility-15	b, c	4
H-16	Histocompatibility-16	b, c	4
H-17	Histocompatibility-17	b, c	—
H-18	Histocompatibility-18	b, c	4
H-19	Histocompatibility-19	b, c	8
H-20	Histocompatibility-20	b, c	4
H-21	Histocompatibility-21	b, c	4
H-22	Histocompatibility-22	b, c	7
H-23	Histocompatibility-23	b, c	3
H-24	Histocompatibility-24	b, c	7
H-25	Histocompatibility-25	b, c	—
H-26	Histocompatibility-26	b, c	—
H-27	Histocompatibility-27	b, c	5
H-28	Histocompatibility-28	b, c	—
H-29	Histocompatibility-29	b, c	8
H-30	Histocompatibility-30	b, c	—
H-31	Histocompatibility-31	a, b, c	17
H-32	Histocompatibility-32	a, b	17
H-33	Histocompatibility-33	a, b	17
H-34	Histocompatibility-34	b, c	—
H-35	Histocompatibility-35	b, c	—
H-36	Histocompatibility-36	b, c	—
H-37	Histocompatibility-37	b, c	—
H-38	Histocompatibility-38	b, c	—
H-39	Histocompatibility-39		17
H-X	Histocompatibility-X	b, c, l	X
H-Y	Histocompatibility-Y	a, b	Y

Differentiation alloantigen loci: A vast number of alloantigens have been identified on cells of lymphoid-tissue lineage. A property of these differentiation antigens (e.g., Thy, Ly, Qa, Tla) is their unequal representation on different classes of lymphoid-tissue cells

TABLE 8.   Genetic profiles of erythrocyte antigens

Chromosome	8					2		
Locus	Ea-1	Ea-2	Ea-3	Ea-4	Ea-5	Ea-6	Ea-7	Ea-8
A	o	b	b	a	a	a	b	+
AKR	o	b	.	a	.	a	b	.
BALB/c	o	b	b	a	.	b	b	—
CBA/J	o	b	.	a	.	a	a	.
C3H/He	o	b	b	.	b	a	a	+
C3H/St	o	b	.	.	a	b	.	.
C57BL/6	o	b	.	.	.	a	b	.
C57BL/10	o	.	b	.	.	.	.	+
C57L	o	b	a	a	.	.	b	.
DBA/2	o	b	.	a	b	b	b	—
NZB	o	b	.	a	.	.	.	+
RF	o	a	.	.	b	b	b	.
RIII	o	a	.	a	b	.	b	.
SJL	o	b	.	a	.	.	b	.
129	o	b	.	a	a	a	a	—

All laboratory stocks tested so far carry the null allele Ea-1o; Ea-1a and Ea-1b are found only in wild mice.

and their detectability by a standard cytotoxic test. The theta or Thy-1 antigen is exclusively expressed on thymocytes and T cells. Ly-antigens describe subsets and determine functions of T and B cells, respectively. In Table 9, selected strain distribution patterns are given for some of the differentiation alloantigen loci. Certain AKR substrains are of Thy-1.2 rather than Thy-1.1 phenotype. This is an example of genetic out-crossing. The CBA/N substrain, which exhibits an X-linked deletion of a mature B-lymphocyte subset can readily be distinguished from other CBA substrains by its non-reactivity with the respective alloantisera (Lyb-3, Lyb-5).

Histocompatibility loci: The antigens controlled by these loci are present on all cells. Several hundred independent histocompatibility loci have been assumed, and these serve as markers of virtually every chromosome (Demant, 1980). Markers that have been defined are the major histocompatibility complex (*H-2*, Chr. 17) and a large number of non-*H-2* loci (partially assigned to certain chromosomes). *H-2* is highly polymorphic and can be demonstrated either through cellular immunity or serologically by antibodies directed against K-, D-, and I-region antigens. The standard haplotypes are defined by private alleles of the K and D end (Table 10). Thus, by antisera directed against the pri-

TABLE 9. *Strain profiles of differentiation alloantigen determining loci*

| Chromosome | 2 | 1 | 9 | | | 4 | 19 | 6 | 6 | 17 | 17 | 17 | 9 | 17 |
Locus	Ly-4	Ly-5	Ly-6	Ly-7	Ly-8	Lyb-2	Lyt-1	Lyt-2	Lyt-3	Qa-1	Qa-2	Qa-3	Thy-1	Tla
A	a	a	a	b	a	b	b	b	b	a	a	a	b	a
AKR	a	a	b	b	b	c	b	a	a	b	b	b	a	b
BALB/cBy	a	a	a	b	a	b	b	b	a	b	b	b	b	c
BALB/cJ	a	a	a	b	a	b	b	b	a	b	a	a	b	c
CBA/J	.	.	.	b	a	a	a	a	.	.	.	.	b	b
C3H/He	a	a	a	b	a	b	a	a	b	.	.	.	b	b
C57BL/6	b	a	b	.	b	b	b	b	b	b	a	a	b	b
C57BL/10	b	a	b	b	.	.	b	b	b	b	a	a	b	c
DBA/2	a	a	.	b	b	a	a	a	b	.	a	a	b	c
RF/J	b	a	.	b	a	c	b	a	a	b	b	b	a	b
SJL/J	a	b	b	b	.	c	b	b	b	a	.	.	b	a
129	a	a	b	b	.	b	b	b	b	b	a	a	b	c

Ly-6 is probably identical to Ly-8.

TABLE 10.   Genetic profiles in histocompatibility-2 region

Haplo-type	Allele K region	Allele D region	Type strain
a	k (23)	d (4)	A; AL; A2G
b	b (33)	b (2)	C57BL/6; C57BL/10; C57L; 129
d	d (31)	d (4)	BALB/c; C57BL/Ks; DBA/2; NZB
f	f (26)	f (9)	RFM/Un
g	d (31)	b (2)	HTG
h	k (23)	b (2)	HTH
j	j (15)	+	I
k	k (23)	k (32)	AKR; CBA/J; C3H/He
p	p (16)	p (22)	BDP; P
q	q (17)	q (30)	AU; BUB; DBA/1
r	r (18)	r (10)	RIII
s	s (19)	s (12)	SJL
u	u (20)	d (4)	PL
z	u (20)	—	NZW

Numbers in parentheses are the antigen private specificity.
$+$:   Possibly a minor variant of $H\text{-}2D^b$.

TABLE 11.   Genetic profiles of alleles of the Igh loci in the Igh-C region

Haplo-type	Type strain	Locus Igh-1	Igh-2	Igh-3	Igh-4	Igh-5	Igh-6
a	BALB/c	a	a	a	a	a	a
b	C57BL/6	b	b	b	b	b	b
c	DBA/2	c	c	c	a	a	—
d	AKR	d	d	d	a	a	—
e	A	e	d	e	a	e	e
f	CE	f	f	f	a	a	—
g	RIII	*	c	g	a	a	—
h	SEA	h	a	a	a	a	—
j	CBA/H	j	a	a	a	a	a

*:   Not all substrains have the same alleles.

vate K and D specificities, most of the haplotypes can be defined. This is of major importance for the monitoring of strains that are congenic at this complex. In practical terms, non-$H\text{-}2$ differences are only demonstrable by histogenetic techniques. Skin grafting therefore is not only used to assure isohistogenicity of an inbred strain. There are several reports indicating that grafts exchanged between substrains behave as allografts.

Allotypes
   Immunoglobulin allotypes: Immunoglobulin allotypes are genet-

ically controlled polymorphic variations of immunoglobulin mole-
cules of the same class. The allotypes of the heavy-, kappa-, and
lambda-chain regions have been designated Igh, Igk, and Igl,
respectively (Green, 1981), and are clustered quite closely on
three different chromosome regions. Six closely linked heavy
chain constant-region loci (Igh-C), designated *Igh-1* to *Igh-6*,
have been mapped on chromosome 12. Thirteen Igh-C haplotypes
are known thus far. Table 11 lists distribution patterns of the
Igh-C loci of the known prototype strains. The two methods most
widely applied for demonstrating immunoglobulin allotypes are
immunodiffusion and passive hemagglutination.

Serum protein allotypes: Under this heading certain proteins are
listed such as hemolytic complement (*Hc*), which is the fifth
component of complement and is either present or lacking. The
trait behaves as a dominant. Another locus is the serum antigenic
substance (*Sas-1*), whose biological function is uncertain. Like
Hc, precipitating antibodies can be raised. The trait is inherited
as an autosomal dominant. The antibodies raised against the
products of these loci are detected by immunochemical methods
such as immunodiffusion. The value of these markers becomes
distinct when comparing, for example, the two congenic strains
B10.D2/oSn and B10.D2/nSn. The former strain, in addition to
the *H-2^d* haplotype, has retained the *Hc^0* allele of DBA/2, whereas
the latter is *Hc^1* as the background strain C57BL/10.

Laboratory Procedures

Coat color

Coat colors are distinguishable by the naked eye, and no special equipment is required to test them. However, age of animals should be taken into consideration when observing coat color, because some coat colors change with age. Coat color phenotypes are generally sufficiently expressed around two weeks after birth, but special care is necessary to distinguish some dilute (d/d) color phenotypes. Young animals usually show darker coat colors than adults, and animals older than 40 weeks have coat colors that turn lighter in appearance.

Six typical coat color phenotypes of adult mice (60–100 days of age) are shown in the Frontispiece.

Dominant alleles cover the appearance of recessive alleles, and the epistatic effect of a locus also masks all phenotypic expression of the hypostatic loci. Test crossing is required to uncover the exact coat color genotypes. DBA strain is usually used for the purpose. Some tester strains for coat colors have been developed (Kondo and Esaki, 1962).

Some examples of test crossing for albino strains with DBA strain are shown in Table 12. Use of the tester strain named ITES, which has genotypes of a/a, b/b, C/C, d/d, and s/s, allows the monitoring of four loci simultaneously.

The crossing test requires several weeks until F_1 progeny have reached the age at which full expression of coat color phenotypes occurs.

Special care should be taken of the F_1 progeny and their mothers used for testing because they themselves could be a cause of genetic contamination. They should therefore be sacrificed as soon

Table 12. Coat color test crossing of albino mouse strains with DBA

DBA genotype		Albino strains and their genotypes		F1 phenotypes and their assumed genotypes
DBA		A		brown
C/C a/a b/b d/d	×	c/c a/a b/b D/D		C/c a/a b/b D/d
		AKR		black
	×	c/c a/a B/B D/D		C/c a/a B/b D/d
		BALB/c		brown-agouti
	×	c/c A/A b/b D/D		C/c A/a b/b D/d
		DDK		black-agouti
	×	c/c A/A B/B D/D		C/c A/a B/b D/d
		HRS		yellow-gray
	×	c/c A/A b/b d/d		C/c A/a b/b d/d

as possible after observation.

Biochemical markers

Electrophoresis is applicable for the detection of all biochemical markers listed in Table 4. Special techniques such as the buffer system, electrode voltage, and stain procedures will be given for each biochemical marker.

1. Sample Preparation

Sample 1:   Plasma

>   The plasma fraction is separated from the erythrocyte fraction of heparinized blood by centrifuging at 3,000 rpm for 15 minutes.

Sample 2:   Blood hemolysates

>   Packed red blood cells (RBCs) of the erythrocyte fraction are washed three times by repeating centrifugation in saline. The RBCs are lysed in a threefold volume of water.

Sample 3:   Kidney homogenates

>   Kidney surgically removed is homogenized in a fivefold volume of deionized water. The homogenates are centrifuged at 15,000 rpm at 0°C for 30 minutes and the supernatant is used.

Sample 4:   Liver homogenates

>   Liver surgically removed is homogenized in a fivefold volume of deionized water. The homogenates are centrifuged 15,000 rpm at 0°C for 30 minutes and the supernatant is used.

Sample 5:   Urine

>   Urine from the mouse is diluted in a twofold volume of deionized water.

All samples should be used promptly after preparation. Samples not in use can be stored in a frozen state below −20°C. They should, however, be used within two months. Frequent thawing and freezing causes reductions in enzyme activities or protein denaturations.

2. *Buffer System*

B1. Acetate EDTA (pH 5.4)
     17.01 g $NaC_2H_3O_2$
     2.48 g EDTA (free acid)
     Bring up to 1 liter with water

B2. Borate (pH 8.6)
     18.5 g boric acid (crystal)
     3.65 g NaOH
     Bring up to 1 liter with water

B3. EDTA sodium acetate (pH 5.6)
     17.01 g $NaC_2H_3O_2 \cdot 3H_2O$
     2.48 g EDTA (tetrasodium salt)
     Bring up to 1 liter with water

B4. 0.01 M Phosphate (pH 6.8)
     3.77 g KH_2PO_4
     8.63 g $Na_2HPO_4 \cdot 12H_2O$
     Bring up to 1 liter with water

B5. 0.2 M Phosphate (pH 6.8)
     27.2 g KH_2PO_4
     Dissolve in 800 ml of water
     Adjust to pH 6.8 with 6 N NaOH
     Bring up to 1 liter with water

B6. 0.05 M Phosphate with 10^{-4} M $CaCl_2$ (pH 7.2)
     74.1 ml 0.5 M Na_2HPO_4
     25.9 ml 0.5 M KH_2PO_4
     Bring up to 1 liter with water
     Add 0.011 g $CaCl_2$ or 0.015 g $CaCl_2 \cdot H_2O$

B7. 0.2 M Phosphate chloride (pH 7.5)
     34.8 g K_2HPO_4
     Dissolve in 900 ml of water
     Adjust to pH 7.5 with concentrated HCl
     Bring up to 1 liter with water

B8. Phosphate-citrate (pH 7.0)

45.4 g Na_2HPO_4
5.9 g citric acid monohydrate
Bring up to 1 liter with water
Heat to boiling to stop microbial growth

B9.   0.05 M Potassium phosphate monobasic (pH 7.0)
(Fisher SO-B-108)

B10.   0.1 M Sodium acetate (pH 5.2)
13.6 g $NaC_2H_3O_2$
Dissolve in 800 ml of water
Adjust to pH 5.2 with glacial acetic acid
Bring up to 1 liter with water

B11.   Tris barbital (pH 8.8)
5.75 g tris
2.45 g barbital
9.81 g sodium barbital
Bring up to 1 liter with water

B12.   Tris borate (pH 8.4)
24.0 g tris
12.0 g boric acid
Bring up to 1 liter with water

B13.   Tris citrate (pH 7.6)
12.1 g tris
Dissolve in 600 ml of water
Adjust to pH 7.6 with 10% citric acid
Bring up to 1 liter with water

B14.   Tris citrate (pH 8.2)
10.5 g tris
3.0 g citric acid
Bring up to 1 liter with water

B15.   Tris citrate (pH 8.3)
16.64 g tris
4.2 g citric acid
Bring up to 1 liter with water

B16. Tris citrate (pH 8.6)
 9.25 g tris
 10.5 g citric acid
 Bring up to 1 liter with water
 Adjust to pH 8.6 with 1 M citric acid

B17. Tris EDTA borate (pH 8.4)
 10.91 g tris
 0.6 g EDTA (disodium salt)
 3.1 g boric acid
 Bring up to 1 liter with water

B18. Tris EDTA borate (pH 9.1)
 10.53 g tris
 0.92 g EDTA (disodium salt)
 0.54 g boric acid
 Bring up to 1 liter with water

B19. Tris glycine (pH 8.5)
 3.0 g tris
 14.4 g glycine (NH_3 free)
 Bring up to 1 liter with water

B20. Tris glycine (pH 8.9)
 5.16 g tris
 3.48 g glycine (NH_3 free)
 Bring up to 1 liter with water

B21. Tris glycine (pH 8.9)
 30.0 g tris
 14.4 g glycine (NH_3 free)
 Bring up to 1 liter with water

B22. Tris HCl (pH 7.0, pH 8.0, pH 9.0)
 24.2 g tris
 Dissolve in 800 ml of water
 Adjust pH with concentrated HCl
 Bring up to 1 liter with water

3. Staining Materials

S1. Alpha-naphthyl butyrate in acetone
  1 g alpha-naphthyl butyrate
  6 ml acetone

S2. 0.5% Amido black in 2% acetic acid

S3. 2% Beta-naphthyl acetate in acetone
  2 g beta-naphthyl acetate
  100 ml acetone

S4. 1% Ferric chloride

S5. Fluorescein diacetate
  1 mg fluorescein diacetate
  0.08 ml acetone to dissolve

S6. Fluorescent stain
  3.0 mg 4-methyl-umbelliferyl acetate
  0.1 ml acetone to dissolve
  5 ml potassium phosphate monobasic pH 7.0 (B9)

S7. Glucose-1-phosphate (100 mg/ml)
  1 g disodium glucose-1-phosphate
  10 ml water

S8. Glucose-1, 6-diphosphate 10 mg/ml

S9. 0.5 M Glucose-6-phosphate

S10. 1% molten agar in distilled water. Cool to 56°C before use.

S11. Hydrogen peroxide (0.03 or 0.06%)

S12. 0.25 M Magnesium acetate $Mg(C_2H_3O_2)_2$

S13. 0.5 M Malate
  2.66 g malic acid in 10 ml water
  Adjust to pH 8.0 with NaOH
  Dilute to 40 ml with water

S14. 0.5% Ponceau S in 5% trichloro-acetic acid

S15. 0.5% Ponceau S in 7.5% trichloro-acetic acid

S16. 0.5% Ponceau S in 10% sulfosalicylic acid
(Helena #5526)

S17. 1% Potassium ferricyanide

4. Electrophoresis Methods
1)   Methods for using cellulose acetate plates
(1)   Alkaline phosphatase-1 (*Akp-1* Chromosome No. 1)

Tissue sample:	Kidney in 5 weight distilled water (Sample 3) 0.6 μl
Buffer system:	Tris citrate; pH 8.2 (B14)
Electrophoresis:	Voltage: 200 volts
	Time: 40 minutes
	Migration: Cathode ($-$) to anode ($+$)
Stain procedure:	(a) 50 mg beta-naphthyl-phosphoric acid
	5 ml distilled water
	(b) 50 mg fast blue RR salt
	10 mg $MgCl_2 \cdot 6H_2O$
	10 mg $MnCl_2 \cdot 4H_2O$
	5 ml tris HCl; pH 9.0 (B22)
	Mix (a) and (b) just before staining
	Incubate at 37°C for 20 minutes
	Fix in 5% acetic acid solution
Type strain:	Akp-1A: C57BL/6
	Akp-1B: C3H/He

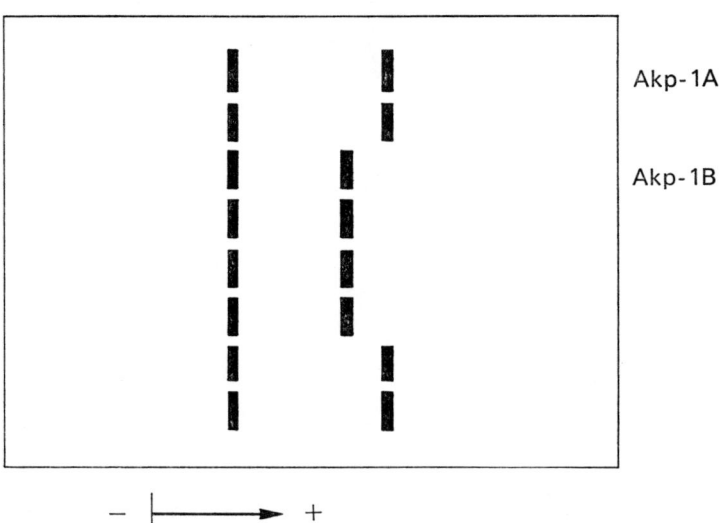

Akp-1A

Akp-1B

$-$ |⟶ $+$

Fig. 5

(2) Carbonic anhydrase-2 (*Car-2* Chromosome No. 3)
  i) Method 1

Tissue sample:	RBCs in 3 volumes distilled water (Sample 2) 0.6 μl
Buffer system:	EDTA sodium acetate; pH 5.6 (B3)
Electrophoresis:	Voltage: 120 volts
	Time: 60 minutes
	Migration: Anode (+) to cathode (−)
Stain procedure:	Apply 0.5% Ponceau S (S16)
	Destain in 5% acetic acid
Type strain:	Car-2A: C57BL/6
	Car-2B: DBA/2

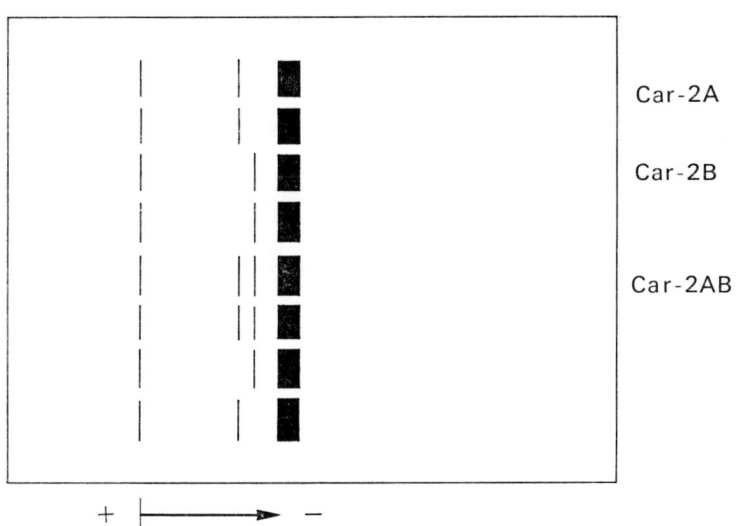

Car-2A

Car-2B

Car-2AB

+ |⟶ −

FIG. 6

ii)   Method 2   (*Car-2*)

Tissue sample: RBCs in equal volume distilled water (Sample 2)

Buffer system: Acetate EDTA; pH 5.4 (B1)
                   Plate 1:8 dilution
                   Electrode 1:8 dilution

Electrophoresis: Voltage: 350 volts
                   Time: 20 minutes
                   Migration: Anode (+) to cathode (−)

Stain procedure: Apply 0.5% Ponceau S (S15)
                   Destain in 2% acetic acid

Type strain: Car-2A: C57BL/6
                   Car-2B: DBA/2

iii) Method 3   *(Car-2)*

Tissue sample:	RBCs in 2 volumes distilled water (Sample 2)
Buffer system:	Acetate EDTA; pH 5.4 (B1)
	Plate 1 : 4 dilution
	Electrode 1 : 4 dilution
Electrophoresis:	Voltage: 250 volts
	Time: 60 minutes
	Migration: Anode $(+)$ to cathode$(-)$
Stain procedure:	Apply 0.5% Ponceau S (S14)
	Destain in 1% acetic acid
Type strain:	Car-2A: C57BL/6
	Car-2B: DBA/2

(3)   Kidney catalase-2 (*Ce-2* Chromosome No. 17)
  i)   Method 1

Tissue sample: Kidney in 10 weight distilled water (Sample
                     3)
                     Let sit 30 minutes
                     Dilute 1: 1 before application
Buffer system: Tris citrate; pH 8.6 (B16)
Electrophoresis: Voltage: 350 volts
                     Time: 15 minutes
                     Migration: Cathode $(-)$ to anode $(+)$
Stain procedure*:   Flood with 0.06% hydrogen peroxide (S11)
                     for 1 minute and rinse with water
                     After blotting apply ferricyanide stain 1: 1
                     (S4 and S17) until bands appear
                     Flood with water
Type strain: Ce-2A: C57BL/6
                     Ce-2B: C3H/He

---

* CAUTION: Cyanide compounds in contact with acids yield lethal vapors ! ! !

ii)   Method 2   (*Ce-2*)

Tissue sample: Kidney in 10 weight distilled water (Sample
                      3)

Buffer system: Tris glycine: pH 8.5 (B19)

Electrophoresis: Voltage: 200 volts
                      Time: 25 minutes
                      Migration: Cathode ($-$) to anode ($+$)

Stain procedure*:   Flood with 0.03% hydrogen peroxide (S11)
                      for 30 seconds and rinse with water
                      Apply ferricyanide stain 1: 1 (S4 and S17)

Type strain: Ce-2A: C57BL/6
                      Ce-2B: C3H/He

---

* CAUTION: Cyanide compounds in contact with acids yield lethal vapors ! ! !

iii)　Method 3　　　(*Ce-2*)

Tissue sample:　　Kidney in 10 weight distilled water (Sample 3)

Buffer system:　　Tris barbital; pH 8.8 (B11)

Electrophoresis:　Voltage: 150 volts

　　　　　　　　Time:　60 minutes

　　　　　　　　Migration: Cathode ($-$) to anode ($+$)

Stain procedure*:　Flood with 0.03% hydrogen peroxide (S11) for 30 seconds and rinse with water

　　　　　　　　After blotting apply ferricyanide stain 1: 1 (S4 and S17) until bands appear

　　　　　　　　Flood with water

Type strain:　　　Ce-2A:　C57BL/6

　　　　　　　　Ce-2B:　C3H/He

---

* CAUTION: Cyanide compounds in contact with acids yield lethal vapors ! ! !

(4) Esterase-1 (*Es-1* Chromosome No.8)

Tissue sample:	Plasma (Sample 1) 0.3 μl
Buffer system:	Phosphate; pH 6.8 (B4)
Electrophoresis:	Voltage: 140 volts
	Time: 30 minutes
	Migration: Cathode ($-$) to anode ($+$)
Stain procedure:	0.5 ml 2% beta-naphthyl acetate (S3)
	50 mg fast blue RR salt
	10 ml phosphate; pH 6.8 (B4)
	Incubate at 37°C until bands appear
	Fix in 5% acetic acid
Type strain:	Es-1A: C57BL/6
	Es-1B: DBA/2

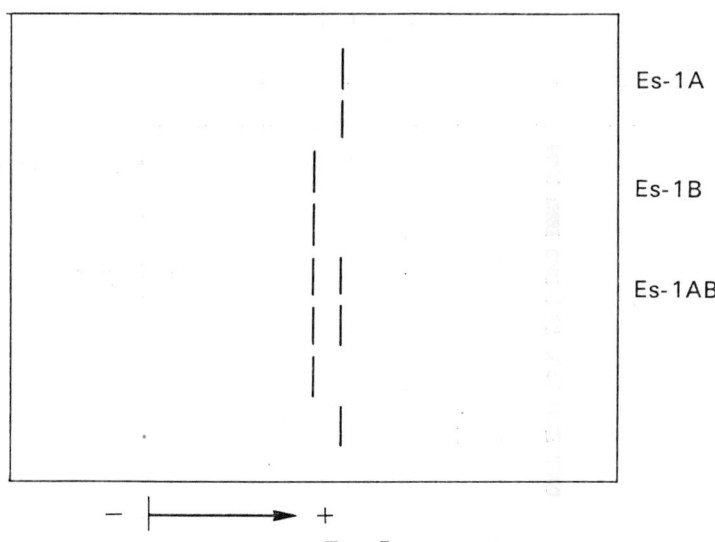

Es-1A

Es-1B

Es-1AB

$-$ |⟶ $+$

Fɪɢ. 7

(5)   Esterase-2 (*Es-2* Chromosome No. 8)

Tissue sample:	Kidney in 5 weight distilled water (Sample 3) 0.6 μl
Buffer system:	Phosphate; pH 6.8 (B4)
Electrophoresis:	Voltage 140 volts
	Time: 30 minutes
	Migration: Cathode $(-)$ to anode$(+)$
Stain procedure:	0.5 ml 2% beta-naphthyl acetate (S3)
	50 mg fast blue RR salt
	10 ml phosphate; pH 6.8 (B4)
	Incubate at 37°C until bands appear
	Fix in 5% acetic acid
Type strain:	Es-2A: KK
	Es-2B: C57BL/6
	Es-2C: PL/J

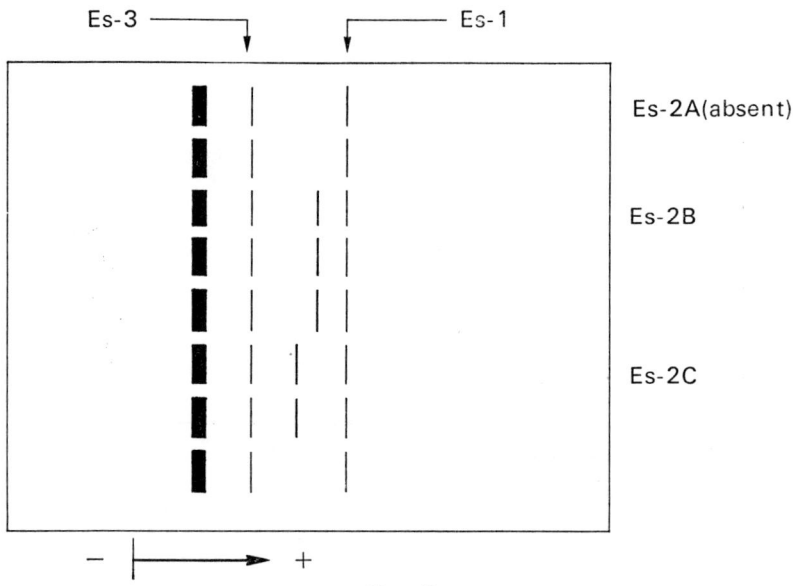

Fig. 8

(6)   Esterase-3 (*Es-3* Chromosome No. 11)

   i)   Method 1

Tissue sample:	Kidney in 5 weight distilled water (Sample 3) 0.6 μl
Buffer system:	Tris EDTA borate; pH 8.4 (B17)
Electrophoresis:	Voltage: 350 volts
	Time: 20 minutes
	Migration: Cathode ($-$) to anode($+$)
Stain procedure:	0.5 ml 2% beta-naphthyl acetate (S3)
	50 mg fast blue SS salt
	10 ml phosphate; pH 6.8 (B4)
	Incubate at 37°C until bands appear
	Fix in 5% acetic acid
Type strain:	Es-3A: C57BL/6
	Es-3B: RF/J
	Es-3C: DBA/2

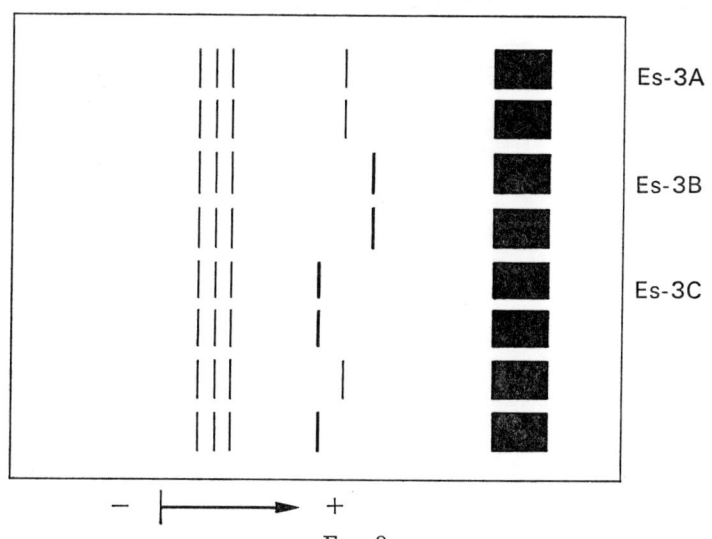

Fig. 9

ii)   Method 2 (*Es-3*)

Tissue sample:	Kidney or liver in 10 weight distilled water (Sample 3 or 4)
Buffer system:	Tris glycine; pH 8.9 (B20)
Electrophoresis:	Voltage: 300 volts
	Time: 25 to 30 minutes
	Migration: Cathode (−) to anode (+)
Stain procedure:	In agar overlay:
	5 ml 1% molten agar (S10)
	3 mg 4-methyl-umbelliferyl acetate dissolved in 0.04 ml acetone
	5 ml phosphate buffer; pH 6.8 (B5)
	Incubate 15 minutes at 37°C
	Read under UV light
Type strain:	Es-3A: C57BL/6
	Es-3B: RF/J
	Es-3C: DBA/2

(7) Esterase-10 (*Es-10* Chromosome No. 14)

Tissue sample:	RBCs in 3 volumes distilled water (Sample 2)
Buffer system:	Tris glycine; pH 8.9 (B20)
Electrophoresis:	Voltage: 300 volts
	Time: 25 minutes
	Migration: Cathode (−) to anode (+)
Stain procedure:	Flood with fluorescent stain (S6)
	Incubate 15 minutes at 37°C
	Read under UV light
Type strain:	Es-10A: C57BL/6
	Es-10B: DBA/2
	Es-10C: BUB/BnJ

(8) Glucose-6-phosphate dehydrogenase-1 (*Gpd-1* Chromosome
   No. 4)
   i) Method 1
   Tissue sample: Kidney in 5 weight distilled water (Sample
                         3) 0.6 μl
   Buffer system: Tris EDTA borate; pH 8.4 (B17)
   Electrophoresis: Voltage: 350 volts
                         Time: 20 to 30 minutes
                         Migration: Cathode ($-$) to anode ($+$)
   Stain procedure: 20 mg glucose-6-phosphate
                         4 mg MTT
                         2 mg PMS
                         8 mg NADP$^+$
                         5 mg $MgCl_2 \cdot 6H_2O$
                         10 ml Tris-HCl; pH 7.0 (B22)
                         Incubate at 37°C until bands appear
                         Fix with 5% acetic acid
   Type strain: Gpd-1A: C57BL/6
                         Gpd-1B: DBA/2
                         Gpd-1C: Several wild mice

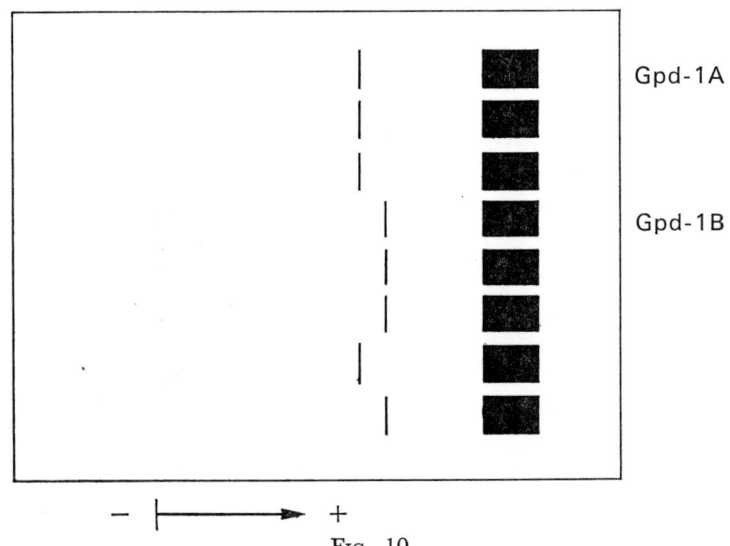

Fig. 10

ii) Method 2 (*Gpd-1*)

Tissue sample:	Liver in 10 weight distilled water (Sample 4) Add 0.04–0.08 ml NADP (10 mg/ml)
Buffer system:	Tris glycine; pH 8.9 (B20)
Electrophoresis:	Voltage: 200 volts
	Time: 1 hour
	Migration: Cathode (−) to anode (+)
Stain procedure:	In agar overlay:
	3 ml 1% molten agar (S10)
	2 ml Tris-HCl; pH 8.0 (B22)
	0.12 ml 0.25 M Mg(C₂H₃O₂)₂ (S12)
	0.12 ml NADP (10 mg/ml)
	0.12 ml 0.008 M PMS
	0.12 ml 0.02 M MTT
	0.24 ml 0.5 M G-6-P
Type strain:	Gpd-1A: C57BL/6
	Gpd-1B: DBA/2
	Gpd-1C: Several wild mice

(9) Glucose-phosphate isomerase-1 (*Gpi-1* Chromosome No. 7)
  i) Method 1

Tissue sample:	Kidney in 5 weight distilled water (Sample 3) Dilute 1:10 with distilled water. 0.3 μl
Buffer system:	Phosphate; pH 6.8 (B4)
Electrophoresis:	Voltage: 160 volts
	Time: 60 minutes
	Migration: Cathode (−) to anode (+)
Stain procedure:	20 mg fructose-6-phosphate
	4 mg MTT
	6 mg PMS
	8 mg NADP+
	20 units G-6-PD
	16 mg $MgCl_2 \cdot 6H_2O$
	10 ml Tris-HCl; pH 8.0 (B22)
	Incubate at 37°C until bands appear
	Fix in 5% acetic acid
Type strain:	Gpi-1A: DBA/2
	Gpi-1B: C57BL/6

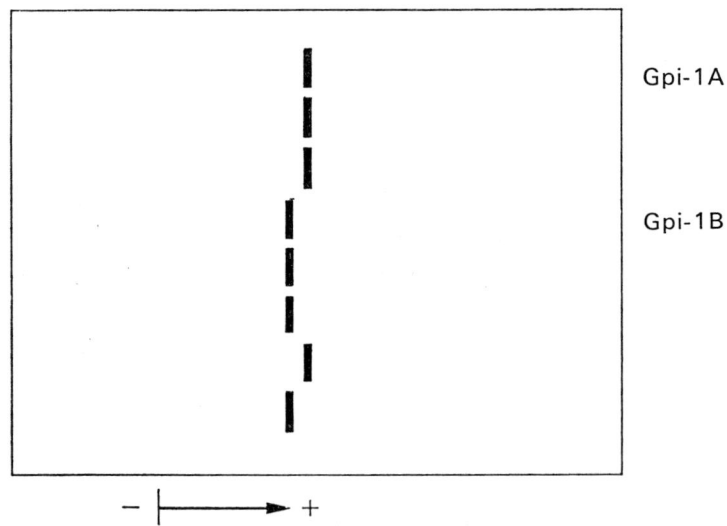

Gpi-1A

Gpi-1B

− ⊢——▶ +

Fig. 11

ii)   Method 2 (*Gpi-1*)

Tissue sample:	RBCs in 3 volumes distilled water (Sample 2) or kidney in 10 weight distilled water (Sample 3) or liver in 10 weight distilled water (Sample 4)
Buffer system:	Tris glycine; pH 8.5 (B19)
Electrophoresis:	Voltage: 200 volts
	Time: 30 minutes
	Migration: Cathode (−) to anode (+)
Stain procedure:	In agar overlay:
	9 ml 1% molten agar (S10)
	2 ml Tris-HCl; pH 8.0 (B22)
	0.12 ml 0.25 M Mg $(C_2H_3O_2)_2$ (S12)
	0.12 ml NADP (10 mg/ml)
	0.12 ml 0.02 M MTT
	0.12 ml 0.008 M PMS
	0.12 ml 10% F-6-P
	0.003 ml G-6-PD (about one enzyme unit)
Type strain:	Gpi-1A: DBA/2
	Gpi-1B: C57BL/6

(10)   Hemoglobin beta-chain (*Hbb* Chromosome No. 7)
  i)   Method 1
   Tissue sample: RBCs in 3 volumes distilled water (Sample
                       2) 0.3 μl
   Buffer system: Tris EDTA borate; pH 8.4 (B17)
   Electrophoresis: Voltage: 350 volts
                       Time: 30 minutes
                       Migration: Cathode (−) to anode (+)
   Stain procedure: Apply 0.5% Ponceau S (S16)
                       Destain in 5% acetic acid
   Type strain: HbbD: DBA/2
                       HbbP: AU/SsJ
                       HbbS: C57BL/6

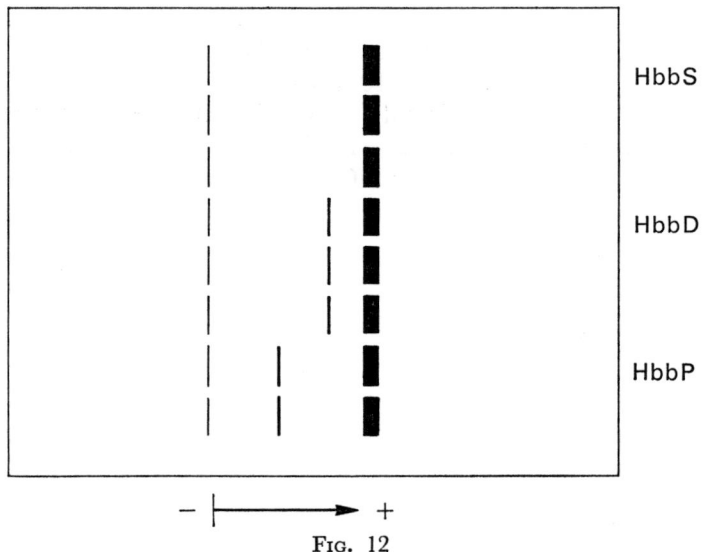

FIG. 12

ii)   Method 2 (*Hbb*)

Tissue sample:	RBCs in 20 volumes distilled water (Sample 2)
Buffer system:	Tris glycine; pH 8.9 (B20)
Electrophoresis:	Voltage: 350 volts
	Time: 12 minutes
	Migration: Cathode (−) to anode (+)
Stain procedure:	0.02 g *o*-dianisidine
	0.5 ml glacial acetic acid
	5.0 ml water
	0.7 ml 3% H_2O_2
	Flood plate until bands turn green
	Rinse with water
Type strain:	HbbD: DBA/2
	HbbP: AU/SsJ
	HbbS: C57BL/6

iii) Method 3 (*Hbb*)

Tissue sample:	RBCs in 3 volumes distilled water
Buffer system:	Tris glycine; pH 8.5 (B19)
Electrophoresis:	Voltage: 200 volts
	Time: 30 minutes
	Migration: Cathode (−) to anode (+)
Stain procedure:	Apply 0.5% Ponceau S (S14)
	Destain in 5% acetic acid

Alkylation for determination of heterozygocity

Tissue sample:	50 μl packed RBCs
	50 μl distilled water
	25 μl toluene
	Mix and centrifuge
	1 volume RBC lysate
	1 volume 1% iodoacetate in 0.05 M phosphate; pH 7.0
	Mix and keep at 4°C overnight
Buffer system:	Tris barbital; pH 8.8 (B11)
Electrophoresis:	Voltage: 200 volts
	Time: 40 minutes
	Migration: Cathode (−) to anode (+)
Stain procedure:	Apply 0.5% Ponceau S (S16)
	Destain in 5% acetic acid

(11) Isocitrate dehydrogenase-1 (*Idh-1* Chromosome No. 1)
  i) Method 1

Tissue sample:	Kidney in 5 weight distilled water (Sample 3) 0.6 μl
Buffer system:	Tris citrate; pH 8.6 (B16)
Electrophoresis:	Voltage: 200 volts
	Time: 40 minutes
	Migration: Cathode (−) to anode (+)
Stain procedure:	4 mg NBT
	3 mg PMS
	4 mg NADP+
	40 mg DL-isocitrate (trisodium salt)
	50 mg $MnCl_2 \cdot 4H_2O$
	10 ml distilled water
	Incubate at 37°C until bands appear
	Fix in 5% acetic acid
Type strain:	Idh-1A: C57BL/6
	Idh-1B: DBA/2

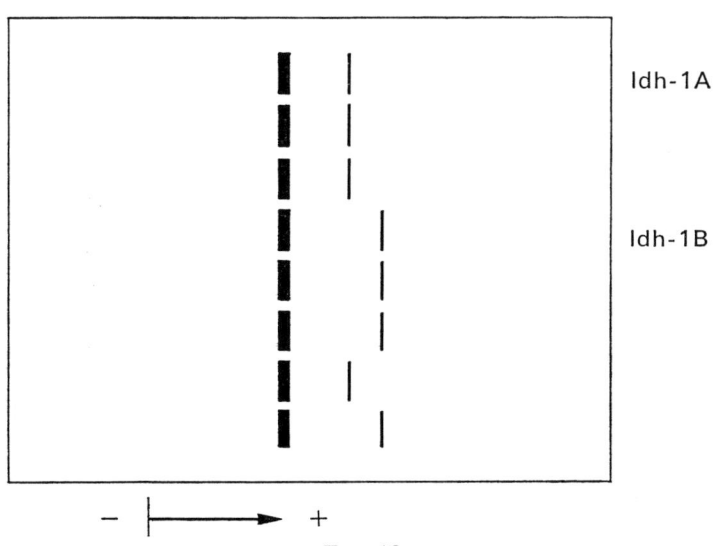

Fig. 13

ii)   Method 2 (*Idh-1*)

Tissue sample: Kidney in 10 weight distilled water (Sample 3)

Buffer system: Phosphate citrate; pH 7.0 (B8)
                         Plate 1: 80 dilution
                         Electrode 1: 10 dilution

Electrophoresis: Voltage: 350 volts
                         Time: 15 minutes
                         Migration: Cathode (−) to anode (+)

Stain procedure: In agar overlay:
                         9 ml 1% molten agar (S10)
                         0.025 ml 0.5 M DL-isocitrate (trisodium salt)
                         0.025 ml 0.07 M $MnCl_2$
                         0.1 ml NADP (10 mg/ml)
                         0.1 ml 0.008 M PMS
                         0.1 ml 0.02 M MTT

Type strain: Idh-1A: C57BL/6
                         Idh-1B: DBA/2

iii)   Method 3 (*Idh-1*)

Tissue sample: Kidney in 10 weight distilled water (Sample
                 3)

Buffer system: Tris glycine; pH 8.5 (B19)

Electrophoresis: Voltage: 200 volts
                   Time: 30 minutes
                   Migration: Cathode (−) to anode (+)

Stain procedure: In agar overlay:
                   3 ml 1% molten agar (S10)
                   2 ml Tris HCl; pH 8.0 (B22)
                   0.04 ml 0.1 м $MnCl_2$
                   0.12 ml NADP (10 mg/ml)
                   0.12 ml 0.20 м MTT
                   0.12 ml 0.008 м PMS
                   0.048 ml 10% isocitrate (trisodium salt)

Type strain: Idh-1A: C57BL/6
               Idh-1B: DBA/2

(12) Lactate dehydrogenase regulator (*Ldr-1* Chromosome No. 6)

Tissue sample: RBCs in 3 volumes distilled water (Sample 2) 0.6 μl

Buffer system: Tris citrate; pH 8.3 (B15)

Electrophoresis: Voltage: 200 volts
Time: 40 minutes
Migration: Cathode (−) to anode (+)

Stain procedure: 1 ml 1 M Na·DL-lactate

$$\left\{\begin{array}{l} 10.6 \text{ ml } 85\% \text{ DL-lactate} \\ 6.1 \text{ g Na}_2\text{CO}_3\cdot\text{H}_2\text{O} \\ \text{Bring up to 100 ml with water} \end{array}\right\}$$

2 mg NBT
1 mg PMS
2 mg NAD+
0.5 ml 0.1 N NaCN
10 ml Tris HCl; pH 7.0 (B22)
Incubate at 37°C until bands appear
Fix in 5% acetic acid

Type strain: Ldr-1A: C57BL/6
Ldr-1B: LP/J, DW/J

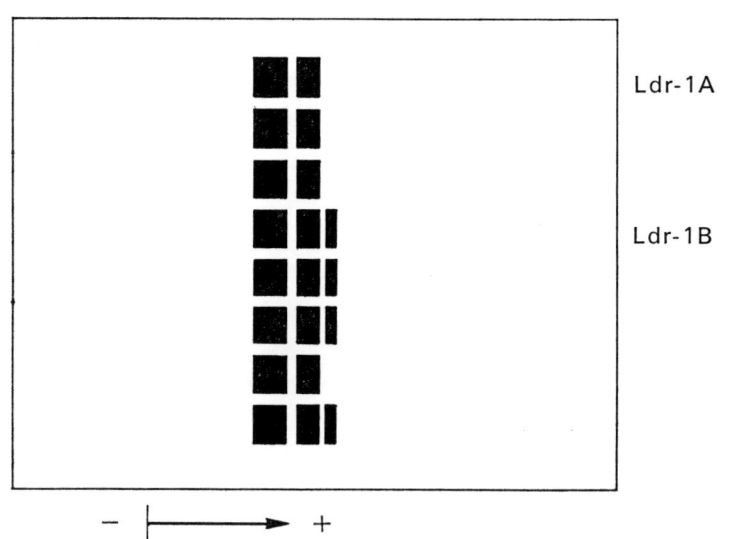

Ldr-1A

Ldr-1B

− |——▶ +

Fɪɢ. 14

(13) Malic enzyme supernatant (*Mod-1* Chromosome No. 9)
  i) Method 1

Tissue sample:	Kidney in 10 weight distilled water (Sample 3)
Buffer system:	Tris citrate; pH 7.6 (B13)
Electrophoresis:	Voltage: 200 volts
	Time: 30 minutes
	Migration: Cathode (—) to anode (+)
Stain procedure:	In agar overlay:
	3 ml 1% molten agar (S10)
	2 ml 0.2 M Tris HCl; pH 8.0 (B22)
	0.12 ml 0.008 M PMS 0.008 M PMS
	0.12 ml 0.02 M MTT
	0.12 ml NADP (10 mg/ml)
	0.12 ml 0.01 M $MnCl_2$
	0.48 ml 0.5 M malate (S13)
	Incubate at 37°C until bands appear
Type strain:	Mod-1A: DBA/2
	Mod-1B: C57BL/6

ii)   Method 2 (*Mod-1*)

Tissue sample:	Kidney in 10 weight distilled water (Sample 3)
Buffer system:	Tris EDTA borate; pH 9.1 (B18)
Electrophoresis:	Voltage: 350 volts
	Time: 12 minutes
	Migration: Cathode ($-$) to anode ($+$)
Stain procedure:	In agar overlay:
	10 ml 1% molten agar (S10)
	0.025 ml 0.07 м $MnCl_2$
	0.5 ml 0.5 м malate (S13)
	0.1 ml NADP (10 mg/ml)
	0.1 ml 0.008 м PMS
	0.1 ml 0.02 м MTT
	Incubate at 37°C until bands appear
Type strain:	Mod-1A: DBA/2
	Mod-1B: C57BL/6

(14)   Major urinary protein (*Mup-1* Chromosome No. 4)
i)   Method 1

Tissue sample:	Urine in same volume distilled water (Sample 5) 0.6 μl for males and 0.9 μl for females
Buffer system:	Tris barbital; pH 8.8 (B11)
Electrophoresis:	Voltage: 160 volts
	Time: 30 minutes
	Migration: Cathode ($-$) to anode ($+$)
Stain procedure:	Apply 5% Ponceau S (S16)
	Destain in 5% acetic acid
Type strain:	Mup-1A: DBA/2
	Mup-1B: C57BL/6

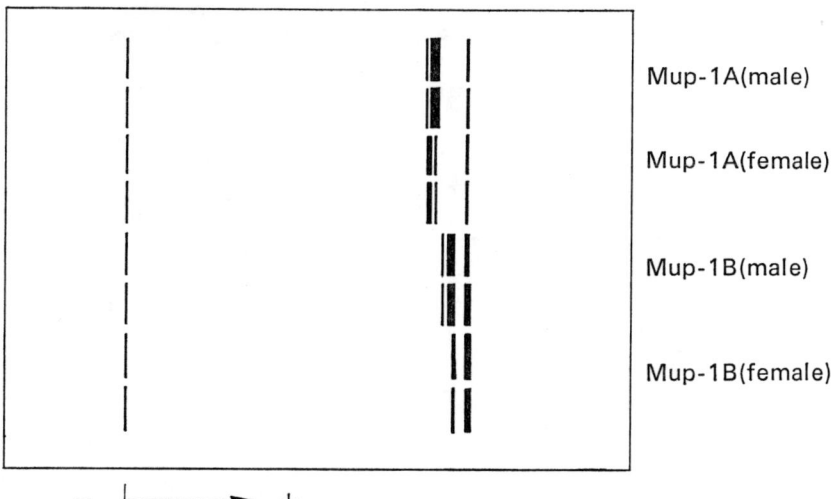

Mup-1A(male)

Mup-1A(female)

Mup-1B(male)

Mup-1B(female)

$-$ |⟶ $+$

Fig. 15

ii)   Method 2 (*Mup-1*)

Tissue sample:	Urine in same volume distilled water (Sample 5)
Buffer system:	Tris glycine; pH 8.5 (B19)
Electrophoresis:	Voltage: 150 volts
	Time: 20 minutes
	Migration: Cathode (−) to anode (+)
Stain procedure:	Apply 5% Ponceau S (S14)
	Destain in 5% acetic acid
Type strain:	Mup-1A: DBA/2
	Mup-1B: C57BL/6

(15)   Peptidase-3 (*Pep-3* Chromosome No. 1)
   i)   Method 1

Tissue sample: Kidney in 10 weight distilled water (Sample 3)

Buffer system: Tris glycine; pH 8.5 (B19)

Electrophoresis: Voltage: 200 volts
                      Time: 25 to 30 minutes
                      Migration: Cathode (−) to anode (+)

Stain procedure: In agar overlay:
                      3 ml molten agar (S10)
                      1 ml phosphate; pH 6.8 (B5)
                      1 ml distilled water
                      0.04 ml Crotalus adamanteus venom (2.5 mg/ml)
                      0.04 ml peroxidase (10 mg/ml)
                      0.04 ml L-leucyl-L-tyrosine (10 mg/ml)
                      0.08 ml 1% *o*-dianisidine
                      Incubate at 37°C until bands appear

Type strain: Pep-3A: C57BL/6
                      Pep-3B: DBA/2
                      Pep-3C: NZB, DD

ii)   Method 2 (*Pep-3*)

Tissue sample: Kidney in 5 weight distilled water (Sample 3) 0.3 μl

Buffer system: Tris EDTA borate; pH 8.4 (B17)

Electrophoresis: Voltage: 300 volts
                   Time: 20 minutes
                   Migration: Cathode (−) to anode (+)

Stain procedure: 20 mg L-leucyl-alanine
                   5 mg MTT
                   5 mg PMS
                   5 mg L-amino acid oxidase
                   5 mg peroxidase
                   10 ml phosphate; pH 6.8 (1 : 15 diluted B4)
                   Incubate at 37°C until bands appear
                   Fix in 5% acetic acid

Type Strain: Pep-3A: C57BL/6
                   Pep-3B: DBA/2
                   Pep-3C: NZB, DD

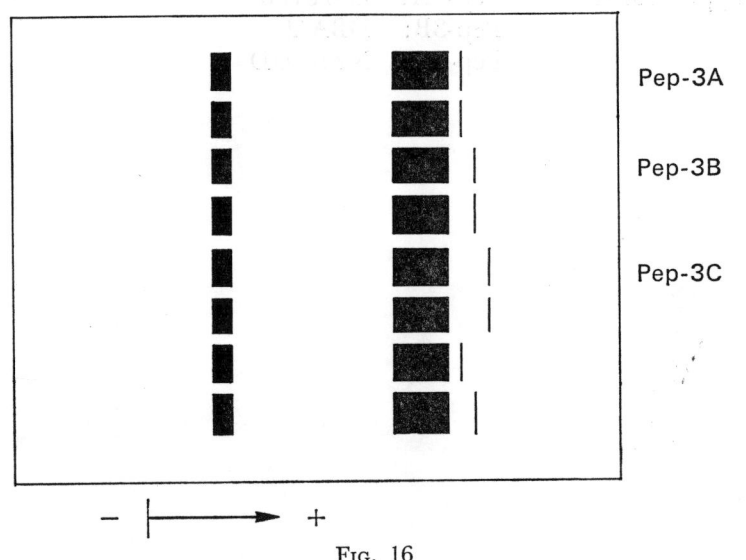

Fig. 16

(16) Phosphoglucomutase-1 (*Pgm-1* Chromosome No. 5)
  i) Method 1

Tissue sample: Kidney in 5 weight distilled water (Sample 3) 0.6 μl

Buffer system: Tris barbital; pH 8.8 (B11)

Electrophoresis   Voltage: 200 volts
           Time: 30 minutes
           Migration: Cathode ($-$) to anode ($+$)

Stain procedure: 40 mg glucose-1-phosphate
           5 mg MTT
           4 mg PMS
           5 mg $NADP^+$
           0.5 mg glucose-1, 6-diphosphate
           20 units glucose-6-phosphate dehydrogenase
           40 mg $MgCl_2 \cdot 6H_2O$
           0.5 ml 1 M NaCN
           10 ml Tris HCl; pH 7.0 (B22)
           Incubate at 37°C until bands appear
           Fix in 5% acetic acid

Type strain: Pgm-1A: C57BL/6
           Pgm-1B: DBA/2

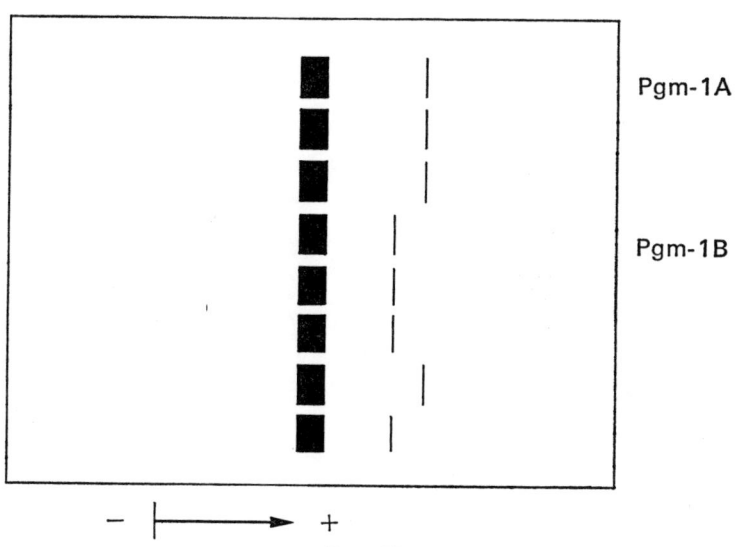

FIG. 17

ii)   Method 2 (*Pgm-1*)

Tissue sample:	Kidney or liver in 10 weight distilled water (Sample 3 or 4) or RBCs (Sample 2)
Buffer system:	Tris glycine; pH 8.5 (B19)
Electrophoresis:	Voltage: 200 volts
	Time: 45 minutes
	Migration: Cathode ($-$) to anode ($+$)
Stain procedure:	In agar overlay:
	3 ml 1% molten agar (S10)
	2 ml Tris HCl; pH 8.0 (B22)
	0.16 ml 0.02 M MTT
	0.16 ml 0.008 M PMS
	0.16 ml NADP (10 mg/ml)
	0.02 ml G-1, 6-DP (S8)
	0.005 ml G-6-PD (300 units/ml)
Type strain:	Pgm-1A: C57BL/6
	Pgm-1B: DBA/2

(17)   Transferrin (*Trf* Chromosome No. 9)
  i)   Method 1

Tissue sample:	Plasma (Sample 1) 0.6 μl
Buffer system:	Tris borate; pH 8.4 (B12)
Electrophoresis:	Voltage: 400 volts
	Time: 30 minutes
	Migration: Cathode (−) to anode (+)
Stain procedure:	Apply 0.5% Ponceau S (S16)
	Destain in 5% acetic acid
Type strain:	TrfA: CBA
	TrfB: C57BL/6

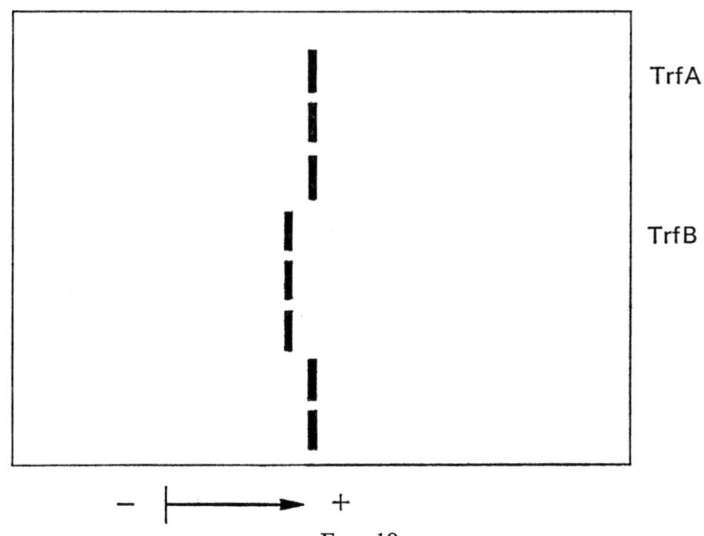

FIG. 18

ii) Method 2 (*Trf*)

Tissue sample: Plasma (Sample 1)

Buffer system: Tris glycine; pH 6.5 (B19)

Electrophoresis: Voltage: 200 volts

Time: 25 minutes

Migration: Cathode (—) to anode (+)

Stain procedure: Apply 0.5% Ponceau S (S14)

Destain in 1% acetic acid

Type strain: TrfA: CBA

TrfB: C57BL/6

2) *Methods for Using Starch Gel*
(1)   Carbonic anhydrase-2 (*Car-2* Chromosome No. 3)

Tissue sample:	RBCs in 3 volumes distilled water (Sample 2)
Buffer system:	Acetate EDTA; pH 5.4 (B1)
	Gel buffer: 1:8 dilution
	Electrode: 1:4 dilution
Electrophoresis:	Voltage: 400 volts
	Time: 3 hours
	Migration: Anode (+) to cathode (−)
Stain procedure:	Apply 0.5% amido black in 2% acetic acid (S2) for 30 seconds
	Destain in 2% acetic acid overnight
Type strain:	Car-2A: C57BL/6
	Car-2B: DBA/2

(2)   Kidney catalase (*Ce-2* Chromosome No. 17)

Tissue sample:	Kidney in 10 weight distilled water (Sample 3)
Buffer system:	Gel buffer: Tris citrate; pH 8.6 (B16) or Tris EDTA borate; pH 9.1 (B18)
	Electrode: Borate; pH 8.6 (B2)
Electrophoresis:	Voltage: 500 volts
	Time: 2 hours
	Migration: Cathode (−) to anode (+)
Stain procedure*:	Flood with 0.03% hydrogen peroxide (S11) for 1 minute
	Rinse with water 2 times. Blot.
	Apply ferricyanide stain 1:1 (S4 and S17) until bands appear
	Flood with water
Type strain:	Ce-2A: C57BL/6
	Ce-2B: C3H/He, AKR

---

* CAUTION: Cyanide compounds in contact with acids yield lethal vapors ! ! !

(3)   Esterase-1 (*Es-1* Chromosome No. 8)*

Tissue sample: Kidney in 10 weight distilled water (Sample 3)

Buffer system: Gel buffer: Tris EDTA borate; pH 9.1
                      (B18)
                      Electrode: Borate; pH 8.6 (B2)

Electrophoresis: Voltage: 450 volts
                      Time: 2 hours
                      Migration: Cathode (−) to anode (+)

Stain procedure: 100 ml phosphate with $CaCl_2$; pH 7.2 (B6)
                      0.5 ml alpha-naphthyl butyrate (S1)
                      0.1 g fast blue BB salt
                      Incubate 60 minutes at 37°C

Type strain: Es-1A: C57BL/6
                      Es-1B: DBA/2

---

* *Es-3* can also be classified on this pad.

(4)   Esterase-3 (*Es-3* Chromosome No. 11)*

Tissue sample:	Kidney in 10 weight distilled water (Sample 3)
Buffer system:	Gel buffer: Tris EDTA borate; pH 9.1 (B18)
	Electrode: Borate; pH 8.6 (B2)
Electrophoresis:	Voltage: 450 volts
	Time: 2 hours
	Migration: Cathode (−) to anode (+)
Stain procedure:	100 ml phosphate with $CaCl_2$; pH 7.2 (B6)
	0.5 ml alpha-naphthyl butyrate (S1)
	0.1 g fast blue BB salt
	Incubate 60 minutes at 37°C
Type strain:	Es-3A: C57BL/6
	Es-3B: RF/J, GR
	Es-3C: DBA/2

---

* *Es-1* can also be classified on this pad.

(5) Esterase-10 (*Es-10* Chromosome No. 14)

Tissue sample:	RBCs in 3 volumes distilled water (Sample 2)
Buffer system:	Gel buffer: Tris EDTA borate; pH 9.1 (B18)
	Electrode: Borate; pH 8.6 (B2)
Electrophoresis:	Voltage: 450 volts
	Time: 2 hours
	Migration: Cathode (−) to anode (+)
Stain procedure:	Incubate 10 minutes at 37°C in sodium acetate; pH 5.2 (B10)
	Blot dry
	Flood with fluorescent stain (S6)
	Read under UV light
Type strain:	Es-10A: C57BL/6
	Es-10B: DBA/2
	Es-10C: BUB/BnJ

(6)   Glucose-6-phosphate dehydrogenase-1 (*Gpd-1* Chromosome No. 4)

Tissue sample:	Kidney in 10 weight distilled water (Sample 3)
	Let sit 2 hours
Buffer system:	Gel buffer: Tris EDTA borate: pH 9.1 (B18)
	Electrode: Borate; pH 8.6 (B2)
Electrophoresis:	Voltage: 450 volts
	Time: 2 hours
	Migration: Cathode ($-$) to anode ($+$)
Stain procedure:	25 ml Tris HCl; pH 8.0 (B22)
	75 ml distilled water
	1.8 ml 0.5 M G-6-P
	1.0 ml 0.25 M Mg $(C_2H_3O_2)_2$ (S12)
	1.0 ml NADP (10 mg/ml)
	1.0 ml 0.008 M PMS
	1.0 ml 0.02 M MTT
	Incubate 45 to 60 minutes at 37°C
Type strain:	Gpd-1A: C57BL/6
	Gpd-1B: DBA/2
	Gpd-1C: Several wild mice

(7) Glucose phosphate isomerase-1 (*Gpi-1* Chromosome No. 7)

Tissue sample:	Kidney in 10 weight distilled water (Sample 3)
	RBCs in 3 volumes distilled water (Sample 2)
Buffer system:	Gel buffer: 3.25 ml phosphate citrate; pH 7.0 (B8)
	Bring up to 250 ml with water
	Electrode: 150 ml phosphate citrate; pH 7.0 (B8)
	Bring up to 1,500 ml with water
Electrophoresis:	Voltage: 450 volts
	Time: 2 hours
	Migration: Cathode (−) to anode (+)
Stain procedure:	In agar overlay:
	1.3 g agar
	120 ml Tris HCl; pH 8.0 (B22)
	6 ml water
	Cool to 55°C and add:
	3.0 ml water
	0.2 ml 0.33 M F-6-P
	0.2 ml 0.25 M $Mg(C_2H_3O_2)_2$ (S12)
	0.2 ml NADP (10 mg/ml)
	0.2 ml 0.008 M PMS
	0.2 ml 0.02 M MTT
	0.03 ml G-6-PD
Type strain:	Gpi-1A: DBA/2
	Gpi-1B: C57BL/6

(8)   Hemoglobin beta-chain (*Hbb* Chromosome No. 7)

Tissue sample: RBCs in 40 volumes distilled water (Sample 2)

Buffer system: Gel buffer: Tris-EDTA-borate; pH 9.1 (B18)
Electrode: Borate; pH 8.6 (B2)

Electrophoresis: Voltage: 450 volts
Time: 60 to 90 minutes
Migration: Cathode (−) to anode (+)

Stain procedure: 0.05 g *o*-dianisidine
1.0 ml water
1.5 ml 3% hydrogen peroxide
Flood pad with stain until bands turn green
Rinse with water

Type strain: HbbD: DBA/2
HbbP: AU/SsJ
HbbS: C57BL/6

(9)   Isocitrate dehydrogenase-1 (*Idh-1* Chromosome No. 1)

Tissue sample: Kidney in 10 weight distilled water (Sample 3)

Buffer system: Gel buffer: 3.25 ml phosphate citrate; pH 7.0 (B8)
        Bring up to 250 ml with water
     Electrode: 150 ml phosphate citrate; pH 7.0 (B8)
        Bring up to 1,500 ml with water

Electrophoresis: Voltage: 450 volts
     Time: 2 hours
     Migration: Cathode (−) to anode (+)

Stain procedure: 25 ml Tris HCl; pH 8.0 (B22)
     75 ml water
     0.5 ml 0.5 M DL-isocitrate (trisodium salt)
     1.0 ml NADP (10 mg/ml)
     1.0 ml 0.008 M PMS
     1.0 ml 0.02 M MTT
     Incubate 15 minutes at 37°C

Type strain: Idh-1A: C57BL/6
     Idh-1B: DBA/2

(10)   Malic enzyme supernatant (*Mod-1* Chromosome No. 9)

Tissue sample:	Kidney in 10 weight distilled water (Sample 3)
Buffer system:	Gel buffer: Tris EDTA borate; pH 9.1 (B18)
	Electrode: Borate; pH 8.6 (B2)
Electrophoresis:	Voltage: 450 volts
	Time: 2 hours
	Migration: Cathode (−) to anode (+)
Stain procedure:	25 ml 0.2 M tris HCl; pH 8.0 (B22)
	75 ml water
	5 ml 0.5 M malate (S13)
	0.25 ml 0.07 M MnCl$_2$
	1.0 ml NADP (10 mg/ml)
	1.0 ml 0.008 M PMS
	1.0 ml 0.02 M MTT
	Incubate 20 to 30 minutes at 37°C
Type strain:	Mod-1A: DBA/2
	Mod-1B: C57BL/6

(11) Major urinary protein (*Mup-1* Chromosome No. 4)

Tissue sample:	Urine (Sample 5)
Buffer system:	Gel buffer: Tris EDTA borate; pH 9.1 (B18)
	Electrode: Borate; pH 8.6 (B2)
Electrophoresis:	Voltage: 450 volts
	Time: 90 minutes
	Migration: Cathode $(-)$ to anode $(+)$
Stain procedure:	Apply 0.5% amido black (S2) for 45 seconds
	Destain in 2% acetic acid overnight
Type strain:	Mup-1A: DBA/2
	Mup-1B: C57BL/6
Note:	Female urine sample can be concentrated 2 times for greater intensity of bands

(12)   Peptidase-3 (*Pep-3* Chromosome No. 1)

Tissue sample:	Kidney in 10 weight distilled water (Sample 3)
Buffer system:	Gel buffer: Tris EDTA borate; pH 9.1 (B18)
	Electrode: Borate; pH 8.6 (B2)
Electrophoresis:	Voltage: 450 volts
	Time: 2 hours
	Migration: Cathode ($-$) to anode ($+$)
Stain procedure:	In agar overlay:
	0.16 g agar
	3 ml 0.2 M phosphate chloride; pH 7.5 (B7)
	8 ml water
	Cool to 55°C and add:
	5 ml 0.2 M phosphate chloride; pH 7.5 (B7)
	0.075 ml 0.1 M $MnCl_2$
	0.15 ml 1% *o*-dianisidine
	0.0032 g Crotalus adamanteus venom
	0.0016 g peroxidase
	0.0064 g L-Leucyl-1-tyrosine
	Incubate 45 to 60 minutes at 37°C
Type strain:	Pep-3A: C57BL/6
	Pep-3B: DBA/2
	Pep-3C: NZB, DD

(13) Phosphoglucomutase-1 (*Pgm-1* Chromosome No. 5)

Tissue sample:	Kidney in 10 weight distilled water (Sample 3)
Buffer system:	Gel buffer: 3.25 ml phosphate citrate; pH 7.0 (B8)
	Bring up to 250 ml with water
	Electrode: 150 ml phosphate citrate; pH 7.0 (B8)
	Bring up to 1,500 ml with water
Electrophoresis:	Voltage: 450 volts
	Time: 2 hours
	Migration: Cathode (−) to anode (+)
Stain procedure:	25 ml tris HCl; pH 8.0 (B22)
	75 ml water
	2.0 ml G-1-P (100 mg/ml) (S7)
	3.0 ml 0.25 M Mg ($C_2H_3O_2$)$_2$ (S12)
	2.0 ml NADP (10 mg/ml)
	1.0 ml 0.008 M PMS
	1.0 ml 0.02 M MTT
	0.05 ml G-6-PD (500 mg/ml)
	Incubate 2 hours at 37°C
Type strain:	Pgm-1A: C57BL/6
	Pgm-1B: DBA/2

(14)   Transferrin (*Trf* Chromosome No. 9)

Tissue sample:	Serum or plasma (Sample 1)
Buffer system:	Gel buffer: Tris EDTA borate; pH 9.1 (B18)
	Electrode: Borate; pH 8.6 (B2)
Electrophoresis:	Voltage: 450 volts
	Time: 3 hours
	Migration: Cathode (−) to anode (+)
Stain procedure:	Apply 0.5% amido black (S2) for 30 seconds
	Rinse 2 times with 2% acetic acid
	Destain overnight in 2% acetic acid
Type strain:	TrfA: CBA
	TrfB: C57BL/6

Immunogenetic markers

No detailed methods are given for the procedures mentioned in this section; rather the principles are discussed, including the advantages and disadvantages and the feasibility of a routine genetic monitoring (possible on a large scale) using each technique. Detailed procedures (tests, serum production, technical equipment, sources) are given in the basic textbook (Carpenter, 1975; Klein, 1975).

Hemagglutination

The hemagglutination technique is applicable for erythrocyte antigens (Ea) and certain private and public H-2 specificities. With this method, erythrocytes are agglutinated by antibodies directed against antigens that are expressed on the target cell. Normally this agglutination does not take place in saline. A developing agent such as polyvinylpyrrolidone (PVP) or dextran must be present in the solution. The standard technique uses PVP as the developing agent. The antiserum is serially diluted in 1% PVP made up in phosphate buffered saline through the addition of 0.1% bovine serum albumin (BSA). Red blood cells are suspended in 0.85% saline at a 2% concentration and added to the serum. After an incubation period and centrifugation, the test is read macroscopically.

The animals to be tested need not be sacrificed. Approximately 80 μl of whole blood from the tail or retroorbital plexus is sufficient, and samples can be taken several times. The test is sensitive and results are obtained rapidly. To set up the test at any laboratory, only standard laboratory equipment is required. Large numbers of animals can be screened by hemagglutination. Since the alleles are codominant (with the exception of *Ea-1ᵉ*), a genetic outcross can be distinguished immediately. If one has to monitor quite a number of H-2 congenic strains on an identical background, serotyping has to be performed. The Ea-antigens also allow for good discrimination between strains and even substrains (Table 8). The proportion of false positive or negative results is rather low.

A major disadvantage of this method is the need for special typing sera or strains for serum production. Some of these sera may be available through the Genetics and Transplantation Biology Brach, NIAID, National Institutes of Health,Bethesda, Maryland, USA, or may be purchased from C.S.E.A.L.-C.N.R.S., Orleans, France. Unfortunately, good hemagglutinating antibodies cannot be induced against all H-2 specificities. In some cases, only cytotoxic antibodies will be raised. The PVP, Type K-60 (average molecular weight 160,000), which is distributed as a 45% aqueous solution, varies from source to source and lot to lot. Either several lots have to be screened, or the PVP concentration has to be altered (0.6–1.2%). In general, the hemagglutination technique will be used for the erythrocyte antigens. For H-2 antigens, the micro-cytotoxicity test seems to be more appropriate, mainly because it saves valuable test serum.

Cytotoxic test

The cytotoxic test—in various modifications—will be used for the demonstration of private and public H-2 antigens as well as differentiation antigens. The antibodies develop their cytotoxic action in the presence of a suitable complement (C′). By means of a vital dye, such as trypan blue or eosin, dead and live cells can be distinguished under a microscope, preferably with phase contrast equipment. Live cells do not stain, whereas dead cells swell and stain. Lymphoid cells are obtained from lymph nodes or thymus by simply tearing them apart in culture medium or from spleen that have been passed through a stainless steel mesh. For a standard test, cells are suspended in culture medium supplemented with fetal calf serum (FCS; 5–10%) or BSA (0.3%) at a concentration of 10×10^6 cells/ml. The viability of untreated cells should be at least 90%. To an appropriately diluted antiserum, an equal volume of cell suspension, usually 10–25 μl, is added (in microtiter plates), mixed, and incubated for 30 minutes at 37°C. Diluted complement (10–25 μl) is then pipetted into each well and after a further incubation period of 30–60 minutes at 37°C the reaction is terminated by placing the plate on ice. Trypan blue (10 μl) is added, and after 3–5 minutes dead and live cells are scored in a hemocytometer. A two-stage modification of this test is more sensitive and is recommended when absorbed antisera are to be used or differentiation antigens to be determined: cell suspension and antiserum are mixed and incubated. Cells are pelleted by centrifugation, and after the supernatant is discarded they are resuspended to the original volume with complement diluted in buffered culture medium. After a second incubation period, the test is scored. Although these tests are highly sensitive, they should be set up in triplicate to overcome problems caused by missing reaction components. Cell and complement controls as well as known positive and negative controls should be included in the test. The cytotoxic index is calculated as follows:

a) $\quad C.I. = 100 \times \dfrac{\% \text{ cells lysed (antibody} + C') - \% \text{ cells lysed (}C' \text{ alone)}}{100 - \% \text{ cells lysed (}C' \text{ alone)}}$

b) $\quad C.I. = 100 \times \dfrac{\% \text{ viable cells (}C' \text{ alone)} - \% \text{ viable cells (antibody} + C')}{\% \text{ viable cells (}C' \text{ alone)}}$

Usually the animals will be sacrificed to obtain thymocytes or large numbers of lymphocytes. For a preliminary testing or for the microcytotoxicity assay, cervical lymph nodes may be removed under slight ether anesthesia. As with the hemagglutination assay, the availability of reagents is a limiting factor. Some antisera for rough H-2 typing and some anti-Ia sera may be obtained from NIAID or C.S.E.A.L., as mentioned above. Alloantisera or monoclonal antibodies against certain differentiation antigens are already commercially available (e.g., Cedarlane Laboratories,

TABLE 13. Donor-recipient combination for production of anti-Ea sera

Specificity	Recipient	Donor	Absorption
Ea-2.1 (R)	(B10 × P.RIII)	B10.RIII	
	C3H	RIII	
	C3H	RFM	
	A.CA	RFM	
Ea-2.2	RIII	C3H	
Ea-4.1	(C3H/HeJ × DBA/2)	C57BL/10	C3H.B10
Ea-4.2	NZB	B10.D2	
Ea-6.1	(C3H/St × DBA/2)	C57BL (leukemia EL4)	
	C3H/St	C3H/He (ascites sarcoma Mc1M)	
Ea-6.2	A	BALB/c (sarcoma Meth A)	
Ea-7.1	C57BL/cd	C3H/He	
Ea-7.2	129	C57BL/10	
	129	B10.A	

TABLE 14.   Donor-recipient combinations for production of anti-H-2 sera

Specificity	Region	Allele	Recipient	Donor
2	D	*b*	(B10.A (5R) × LP.RIII)	B10
4	D	*d*	(B10.AKM × 129)	B10.A
9	D	*f*	(B10.D × SJL)	B10.M
12	D	*s*	(B10 × DBA/1)	DA
15	K	*j*	(B10.A (2R) × A.CA)	B10.WB
16	K	*p*	(A × B10.S)	B10.P
17	K	*q*	(DBA/2 × B10.AKM)	B10.Q
18	K	*r*	(B10 × C3H)	B10.RIII (71NS)
19	K	*s*	(B10.A × DA)	B10.S
20	K	*u*	(DBA/2 × B10.BR)	B10.PL
22	D	*p*	(A.AL × B10.F (13R))	B10.P
23	K	*k*	(B10 × LP.RIII)	B10.A (2R)
26	K	*f*	(A × B10.S)	A.TRF-4
30	D	*q*	(B10.A × LP.RIII)	B10.AKM
31	K	*d*	(B10 × A)	B10.D2
32	D	*k*	(B10.A (2R) × C3H.SW)	C3H
33	K	*b*	(B10.D2 × A)	B10.A (5R)

TABLE 15.   Donor-recipient combination for production of antisera
to define differentiation antigen

Specificity	Recipient	Donor	Absorption
Ly-4.1	(B6.C-H-2d × CXB6)	CXBD	
Ly-4.2	(SWR × BALB/c)	B10.D2	
Ly-5.1	SJL	A.SW	SJL
	(SJL × DA)	A.SW	
Ly-5.2	(B6-H-2k × A.SW)	SJL	A (RADA 1 leukemia)
	(A.SW × B6-Tlaa)	SJL	
Lyb-2.1	(C3H.I × C57BL/6)	I/St (ascites tumor I.29)	C57BL/6
Lyb-2.2	(SJL/J × C/J)	A.SW	HSFS/N
Lyb-2.3	(C3H/An × BALB/c)	CE/J	PL/J and HSFS/N
Lyt-1.1	(BALB/c × C57BL/6)	B6-Ly-1a	
Lyt-1.2	C3H/An	CE	
	C3H/An	C3H.CE/Lyt-1b	
Lyt-2.1	B6-H-2k	CE	
	B6-H-2k	B6-H-2k.CE-Lyt-2a	
Lyt-2.2	(C3H/An × B6-Lyt-2a)	C57BL/6	
	(C3H × BDP)	B10.BR	
Lyt-3.1	(CBA × SJL/J)	C58	
Lyt-3.2	C58	CE	
	C58	C58.CE-Lyt-3b	
	(AKR × C58)	CE	

Ltd. of Canada). Recipient-donor combinations for the production of specific antisera are listed in Tables 13, 14, and 15.

The above outlined tests give rather reliable results and should be applied in establishing the genetic profile of a strain. However, 30 to 75 μl antiserum dilution (triplicates) is needed per serum per animal. For screening procedures, the major task in genetic quality assurance, the microcytotoxicity test using Terasaki trays is more feasible. Trays holding 1 μl of appropriately diluted antiserum per well can be prepared in advance and kept frozen until needed. The test protocol is identical to the one-stage cytotoxicity test, except that 1 μl of cell suspension containing 2,000 cells, 5 μl of complement, and 3 μl of either eosin or trypan blue will be added. Addition of 5 μl buffered formaldehyde (pH 7.3–7.4) will preserve determinations at 4°C for four days. The reactions of the cell-serum-complement-dye mixture are read with an inverted phase contrast microscope using a 16$\times$ objective and 10$\times$ eyepieces.

A major disadvantage of the cytotoxic tests is the need for a high-quality complement source. For H-2 typing, the use of fresh or frozen guinea pig C', tested for low background cytotoxicity, is usually sufficient; however, in our experience with rabbit C' the anti-H-2 sera may be used at a higher dilution, thus saving typing reagents. For most differentiation antigens, rabbit C' is a must. However, most rabbit sera have a high background cytotoxicity. On average the serum of one rabbit out of ten is suitable. There are several commercial sources for guinea pig and rabbit complement, either as standard C' or as C' with low background cytotoxicity.*

By means of defined antisera, not only can the genetic profile for alloantigen markers be established, but also in the case of an outcross the contaminating strain may be identified. If the monitoring laboratory has no access to defined antisera, it can make use of polyvalent antisera (Festing and Totman, 1980) obtained by injecting lymphocytes pooled from several different strains that are maintained at the same location into each single inbred strain. When applying the microcytotoxicity test as de-

---

* Cedarlane Laboratories Ltd., 55168-th Line R.R. 2, Hornby, Ontario, Canada LOP 1EO, Telephone (416) 878–7800

Pel-Freeze Biologicals, P.O. Box 68, Rogers, Arkansas 72756, U.S.A., Telephone (501) 636–2244

scribed, individuals of the antiserum producing strain should *not* react with this serum unless an outcross or a mutation has occurred. The results, however, do not indicate the source of a detected alteration, as is the case when defined monospecific antisera are used.

Since these antisera mostly detect MHC antigens, false negative results have to be taken into consideration.

Since the availability of defined alloantisera is limited, their use should be restricted to the monitoring of the foundation colonies, whereas the polyvalent antiserum is well suited for a large-scale testing of pedigreed expansion colonies and, mainly, production colonies to assure their integrity, if the strains kept together carry different MHC haplotypes.

Skin grafting

Isohistogenicity is a prerequisite for inbred strains. The permanent acceptance of grafts from individuals of the same strain defines whether a strain is isohistogenic or not. To detect H-gene differences, caused by either mutation or outcrossing, skin grafting is one of the most sensitive and widely used techniques. The several hundred H genes monitored by this method serve as markers of virtually every chromosome. So far this monitoring procedure cannot be replaced by any serological or other technique. Skin from the trunk, pinna, or tail may be used as a graft and transplanted onto the lateral chest wall. The procedure is laborious as there are several pre- and postoperative steps, such as hair clipping and bandaging. The graft bed is prepared either leaving the Musculus panniculus carnosus intact (which gives the best results due to better vascularization) or by excising a circular piece of full-thickness skin. The graft is placed in its graft bed, care being taken that the hair growth of the graft (when trunk skin is used) is opposite to that of the host. The grafted tissue is fixed in place by a wound dressing. The rate of technical failure ranges from 5% to 20%.

For routine monitoring, the technique by Bailey and Usama (1960) using orthotopic tail skin grafts is more advisable, since the technique is simple and permits the grafting of a large number of animals within a short period of time. Grafts (7–8 mm × 2–3 mm in size) are sliced from the tail with a scalpel sufficiently thin, so that there is no severe hemorrhage. Reciprocal grafts (Fig. 19) are immediately exchanged and fitted into their graft bed, so that the hair growth is reverse. The graft is fixed with liquid surgical dressing and secured by a glass tube that is slipped onto the tail and fixed with tape. There is a 5 to 10% rate of technical failure. Since the grafts are small, weaker H differences (due to mutation) are more likely to be detected.

Compatible grafts heal in, first (after 6–8 days) becoming swollen and pink due to a hypervascularity. The upper layer of the epidermis may peel off in one piece (ghost graft). Incompatible grafts undergo destruction. The rejection may be either acute (within 3 weeks) or chronic (after 3 weeks). Acute rejection shows all signs of inflammation, like hemorrhage, intense swelling, and a change in color to dark red. Due to necrosis, the graft becomes wet and then transforms into a scablike mass which finally sloughs off. The signs of chronic rejection are a gradual hair loss, scaliness, and cicatrization. Rejection may be complete, leaving a scar, or incomplete, followed by recovery of the graft (the graft is said to have "undergone crisis"). If there is any doubt about the success of graft acceptance regrafting is a must. In case of true incompatibility (and not technical failure), the second set graft will

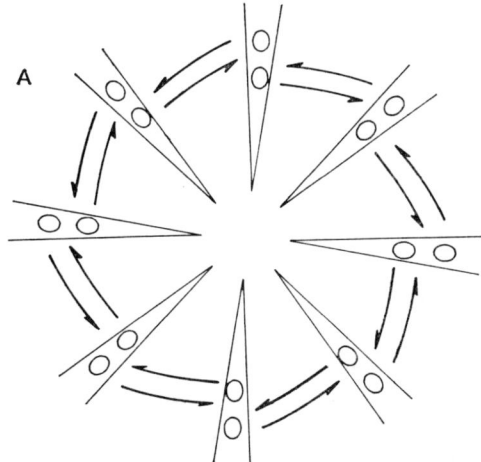

Fig. 19. (A) Reciprocal circle system of skin grafting for detection of H mutations. The large V's represent mouse tails, the ovals within the V's represent grafts, and arrows indicate direction of grafting.

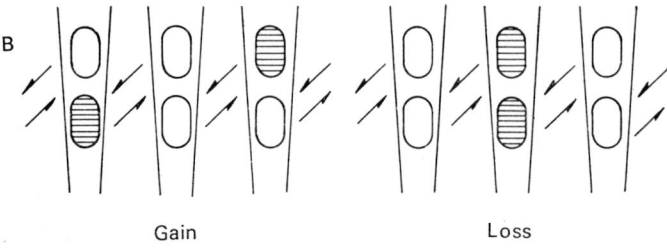

Gain   Loss

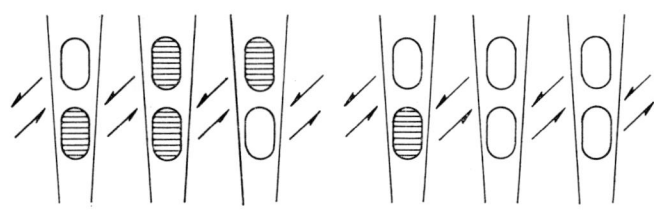

Gain and loss   Indeterminate

(B) Different patterns of graft rejection expected in segments of the reciprocal circle when H mutation occurs. Shaded ovals represent rejected grafts; empty ovals represent permanently accepted grafts.

(Reproduced with permission from D.W. Bailey and H.I. Kohn: "Inherited histocompatibility changes in progeny of irradiated and unirradiated inbred mice." *Genet. Res. Camb.* 6: 330–340, © 1965 Cambridge University Press).

be rejected in a hastened and more pronounced manner. Genetic outcrosses generally cause the grafts to be rejected in acute fashion and are thus immediately detected. Every other generation of grafts should be exchanged in a reciprocal circle within individuals of the foundation colonies, thus calling for one transplantation cycle per year per strain. The tail grafting technique is feasible and easy to perform. The only disadvantages are the long observation period (at least 100 days), the relatively large space needed for holding the animals, and the relatively high percentage of technical failures.

Immunodiffusion

Of the major methods for the determination of immunoglobulin and serum protein allotypes, double immunodiffusion (Ouchterlony analysis) is the most widely applied and feasible technique. Alternative methods that are in use are passive hemagglutination and radioimmunoassays. For immunodiffusion, precipitating antibodies are to be induced: Either normal serum or protein components will serve as antigen for the induction of antibodies against Hc[1] and Sas-1[a]. For the induction of anti-allotype Igh-1 and Igh-1 through Igh-6 serum, respectively, immunoglobulins from animals which contain antibodies directed against some tissue (e.g., anti H-2) or an externally created antigen-antibody complex (Bordetella pertussis, Proteus mirabilis) are to be used as antigen. The choice of the recipient mouse strain strongly influences the success of the anti-allotype immunization. The test is regularly run on a barbital-buffered agar (1%) on a microscopic slide or in tissue culture plates. Circular basins are cut into the agar in a hexagonal pattern. (Agar gel cutters with a pattern for a series of "six-shooter" comparative analyses on slides are commercially available).

A central well contains the typing serum. The unknown serum samples are poured into the peripheral wells. Precipitation lines will form at room temperature within 3–6 hours for Igh-C, 18–24 hours for Hc, and 48 hours for Sas-1. For documentation, these precipitation lines may be permanently stained by a protein stain (buffalo black, Ponceau S) or photographed. The determination of Igh-1 (γ 2a-chain) allotype is sufficient for genetic monitoring purposes, as it allows for a discrimination of 9 codominantly expressed alleles (Table 6) in laboratory mice, and since there is almost no recombination reported within the Igh-C complex. Hc and Sas-1 code for presence or absence of the antigen. The presence of these antigens is inherited as an autosomal dominant. Thus, heterozygotes are phenotypically indistinguishable from homozygous carriers. In doubtful cases F_1 hybrids between the individuals in question and known non-carriers will be decisive.

Immunodiffusion as a test method is feasible and can be established at any laboratory without difficulty. Even with little experience, results can be clearly determined. Although this method is not as sensitive for the recognition of allotypic determinants on immunoglobulin heavy chains as passive hemagglutination or radioimmunoassay, the test is quick and allows characterization of large numbers of samples within a short period of time. Since only a small quantity of serum is required for this test, animals need not be sacrificed. The need for special typing sera is somewhat disadvantageous. However, sera directed against the serum protein allotypes can easily be prepared. Mouse anti-immunoglobulin allotype sera are more difficult and laborious to produce. For detailed protocols and donor-recipient combinations, see Herzenberg and Herzenberg (1978).

Data Analysis

Analyses of monitoring results using qualitative characters are performed by comparing the genotype on each locus of the test strain with the genetic profile.

A computer program for data analysis is given in the Appendix. This program includes genetic profiles on 25 loci of 20 inbred strains of mice. With this program, a computer will display the analyzed data, as shown in Table 16, when we input the genotype on each locus of the test strain and the standard strain name with which we want to compare it.

In the upper part of Table 16, the test strain is compared with the standard AKR strain. We can see that all genotypes of each locus were identical, as shown by the dashes in the bottom line. On the other hand, when this test strain was compared with the BALB/c strain, genotypes in 12 loci were different in each strain as shown by the question marks in the lower part of Table 16.

There is no trouble when all genotypes of the test strain fit the genetic profile. However, we must go to the next step of data analysis when some genotypes of the test strain do not fit the profile. Disagreements between test strain results and genetic profile can be caused by (1) genetic contamination, (2) division of subline before complete inbreeding, and (3) occurrence of mutation. Some symptoms of doubtful genotypes, the reason for their existence, and countermeasures to be taken are shown in Fig. 20.

The existence of heterozygosity in many loci suggests that there was genetic contamination (unplanned mating between different strains) in recent generations: the contaminated strain must be destroyed and replaced with the defined strain. When heterozygosity is observed in only one locus, it is considered that a mutation has occured in recent generations. In this case, we can take either

TABLE 16. Analysis of genetic monitoring results

a) Test strain vs. AKR

Locus	Idh-1	Pep-3	Akp-1	Hc	a	Car-2	Mup-1	b	Gpd-1	Pgm-1	Ldr-1	Gpi-1	p
AKR	b	b	b	o	a	a	a	+	b	a	a	a	+
Test str.	b	b	b	o	a	a	a	+	b	a	a	a	+
Diff.	—	—	—	—	—	—	—	—	—	—	—	—	—

Locus	c	Hbb	Es-1	Es-2	Thy-1	d	Mod-1	Trf	Es-3	Es-10	H2K	H2D
AKR	c	d	b	b	a	+	b	b	c	b	k	k
Test str.	c	d	b	b	a	+	b	b	c	b	k	k
Diff.	—	—	—	—	—	—	—	—	—	—	—	—

b) Test strain vs. BALB/c

Locus	Idh-1	Pep-3	Akp-1	Hc	a	Car-2	Mup-1	b	Gpd-1	Pgm-1	Ldr-1	Gpi-1	p
BALB/c	a	a	b	l	+	b	a	b	b	a	a	a	+
Test str.	b	b	b	o	a	a	a	+	b	a	a	a	+
Diff.	?	?	—	?	?	?	—	?	—	—	—	—	—

Locus	c	Hbb	Es-1	Es-2	Thy-1	d	Mod-1	Trf	Es-3	Es-10	H2K	H2D
BALB/c	c	d	b	b	b	+	b	b	c	b	d	d
Test str.	c	d	b	b	a	+	b	b	c	b	k	k
Diff.	—	—	—	—	?	—	—	—	—	—	?	?

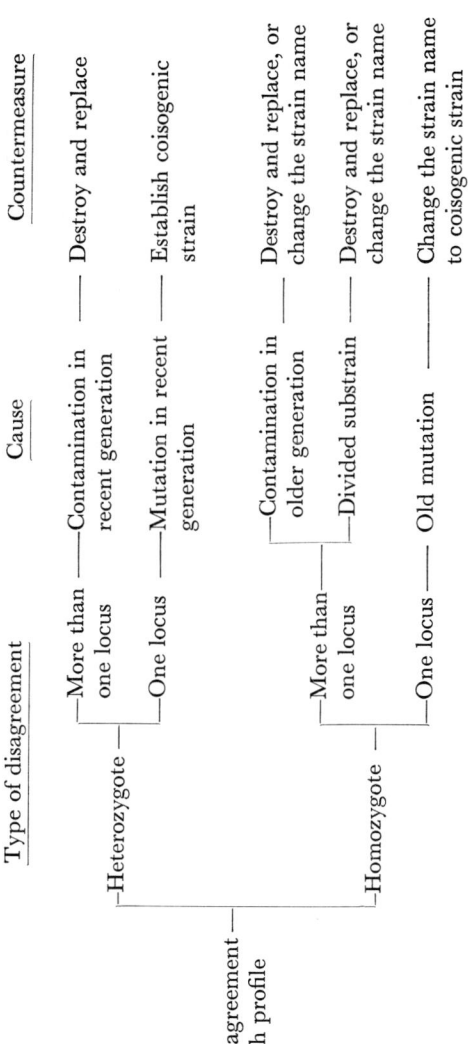

FIG. 20 Types and causes of disagreement with the genetic profile and countermeasures to be taken.

of two measures: 1) destruction of the test strain or 2) establishment of a new widow strain from the test strain.

When all loci observed are homozygous but genotypes at more than one locus are different from those of the standard strain, two possibilities can be considered. One of these is genetic contamination in an older generation. When genetic contamination occurs, many loci will show heterozygosity for several generations. However, if inbreeding was conducted after contamination, some of these heterozygous loci would be fixed randomly to the genotype of the original strain and others would be fixed to genotypes different from the standard. It is clear that such a strain cannot be used in research because it is not the same as the original strain.

The second reason for the existence of homozygotes that are different from the profile at many loci is the division of the branch strain from the main line at an early stage of inbreeding. When this occurs, the strain must be destroyed and replaced with the defined strain, or the strain name must be changed if we want to maintain it as a new strain.

The last case is one where genotypes differ from the profile at only one locus and it is homozygous. An example is when a mutation occurs in an older generation and the allele of this locus is fixed to the mutant gene. When this occurs, the strain name should be changed to that of a new coisogenic strain.

Appendix

```
10 REM *** GENETIC MONITORING ***
20 CONSOLE C40,S0,24:CLR
110 PRINT "    **********************************"
120 PRINT "    *                                *"
130 PRINT "    *   Strain Identification        *"
140 PRINT "    *         Program                *"
150 PRINT "    *                                *"
160 PRINT "    *     using Sharp mz-80B          *"
170 PRINT "    *          computer              *"
180 PRINT "    *                                *"
190 PRINT "    *         Written by             *"
200 PRINT "    *      Kozaburo ESAKI            *"
210 PRINT "    *                                *"
220 PRINT "    **********************************"
230 PRINT:PRINT "Please hit any key"
240 Z$="":GET Z$:IF Z$="" GOTO 240
250 REM *** CREATION OF DIMENSION ***
260 L=20:N=25
270 DIM ST$(L),LO$(4,N),A$(4)
280 RESTORE 9010
290 FOR K=1 TO L
300 READ ST$(K)
310 ST$(K)=LEFT$(ST$(K)+"          ",10)
320 NEXT K
330 RESTORE 9050
340 FOR J=1 TO N
350 READ LO$(1,J)
360 LO$(1,J)=LEFT$(LO$(1,J)+"     ",5)
370 NEXT J
380 PRINT CHR$(6);
390 PRINT "You can compare with following strains:":PRINT
400 FOR K=1 TO L
410 PRINT "     ";ST$(K);
420 IF K/2=INT(K/2) THEN PRINT:PRINT
430 NEXT K: PRINT
440 PRINT "Please hit any key"
450 Z$="":GET Z$:IF Z$="" GOTO 450
460 PRINT CHR$(6);
470 PRINT "and on following loci:":PRINT
480 PRINT "    Idh-1   Akp-1   Pep-3   Hc      a":PRINT
490 PRINT "    Car-2   Mup-1   b       Gpd-1   Pgm-1":PRINT
500 PRINT "    Ldr-1   Gpi-1   p       c       Hbb":PRINT
510 PRINT "    Es-1    Es-2    Thy-1   d       Mod-1":PRINT
520 PRINT "    Trf     Es-3    Es-10   H-2K    H-2D":PRINT:PRINT
530 PRINT "Please hit any key"
540 Z$="":GET Z$:IF Z$="" GOTO 540
```

```
1210 REM ***** DATA INPUT *****
1220 PRINT CHR$(6);
1230 PRINT "Please input data of test strain":PRINT
1240 FOR J=1 TO N
1250 IF J/2=INT(J/2) THEN PRINT TAB(20);CHR$(2);
1260 PRINT LO$(1,J);"=";
1270 INPUT LO$(3,J)
1280 LO$(3,J)=LEFT$(LO$(3,J)+"      ",5)
1290 NEXT J
1300 PRINT CHR$(6);
1310 PRINT "Revision of data":PRINT
1320 FOR J=1 TO N
1330 IF J/2=INT(J/2) THEN PRINT TAB(20);
1340 IF J<10 THEN PRINT CHR$(3);
1350 PRINT J;". ";LO$(1,J);"= ";LO$(3,J);
1360 IF J/2=INT(J/2) THEN PRINT
1370 NEXT J:CONSOLE S20,24
1380 PRINT "  Input No. for correction"
1390 PRINT "      If non, input ┌0┘";
1400 INPUT X
1410 IF X>N THEN PRINT CHR$(6);"Please again":GOTO 1380
1420 CONSOLE S0,24
1430 IF X=0 GOTO 1510
1440 PRINT LO$(1,X);"=";:INPUT LO$(3,X)
1450 LO$(3,X)=LEFT$(LO$(3,X)+"      ",5)
1460 GOTO 1300
1510 REM *** COMPARE WITH STANDARD ***
1520 PRINT "Compare with standard strain":PRINT
1530 FOR K=1 TO L
1540 IF K/2=INT(K/2) THEN PRINT TAB(20);
1550 IF K<10 THEN PRINT CHR$(3);
1560 PRINT K;". ";ST$(K),
1570 IF K/2=INT(K/2) THEN PRINT
1580 NEXT K:CONSOLE S22,24:PRINT
1590 INPUT "Input strain No. ";X
1600 IF X>L THEN PRINT CHR$(6):GOTO 1590
1610 CONSOLE S0,24
1620 GOSUB 2500
1630 FOR J=1 TO N
1640 READ LO$(2,J)
1650 LO$(2,J)=LEFT$(LO$(2,J)+"      ",5)
1660 NEXT J
1670 FOR J=1 TO N
1680 IF LO$(2,J)=LO$(3,J) THEN LO$(4,J)="-      ":GOTO 1700
1690 LO$(4,J)="?      "
1700 NEXT J
```

```
1810 REM ***** PRINT OF RESULTS *****
1820 CONSOLE C80,S0,24
1830 A$(1)="Locus      ":A$(2)=ST$(X):A$(3)="Test str. ":A$(4)="Diff.
1840 PRINT "Test strain vs ";ST$(X)
1850 PRINT STRING$("-",73)
1860 FOR I=1 TO 4
1870 PRINT A$(I);
1880 FOR J=1 TO 13
1890 PRINT LO$(I,J);
1900 NEXT J:PRINT
1910 IF (I=1)+(I=3) THEN PRINT STRING$("-",73)
1920 NEXT I
1930 PRINT STRING$("-",73):PRINT:PRINT
1940 PRINT STRING$("-",70)
1950 FOR I=1 TO 4
1960 PRINT A$(I);
1970 FOR J=14 TO N
1980 PRINT LO$(I,J);
1990 NEXT J:PRINT
2000 IF (I=1)+(I=3) THEN PRINT STRING$("-",70)
2010 NEXT I
2020 PRINT STRING$("-",70)
2030 CONSOLE S20,24
2040 PRINT "Do you need a hard copy ?":PRINT
2050 PRINT "         Input ┌Y┘ for yes, or"
2060 INPUT "               ┌N┘ for no.    ";Z$
2070 IF Z$="Y" THEN PRINT CHR$(6):COPY/P1
2080 PRINT:PRINT "Do you compare with another standard strain ?":PRINT
2090 PRINT "         Input ┌Y┘ for yes, or"
2100 INPUT "               ┌N┘ for no.    ";Z$
2110 CONSOLE C40,S0,24
2120 IF Z$="N" GOTO 2150
2130 GOTO 1510
2150 PRINT "Have you data of other test strain ? ":PRINT
2160 PRINT "         INPUT ┌Y┘ for yes, or":PRINT
2170 INPUT "               ┌N┘ for no.    ";Z$
2180 IF Z$="N" GOTO 9990
2190 FOR I=2 TO 4
2200 A$(I)=""
2210 FOR J=1 TO N
2220 LO$(I,J)=""
2230 NEXT J:NEXT I
2240 GOTO 1210
```

```
2500 REM *** RESTORE SUBROUTIN ***
2510 IF X=1 THEN RESTORE 9110
2520 IF X=2 THEN RESTORE 9120
2530 IF X=3 THEN RESTORE 9130
2540 IF X=4 THEN RESTORE 9140
2550 IF X=5 THEN RESTORE 9150
2560 IF X=6 THEN RESTORE 9160
2570 IF X=7 THEN RESTORE 9170
2580 IF X=8 THEN RESTORE 9180
2590 IF X=9 THEN RESTORE 9190
2600 IF X=10 THEN RESTORE 9200
2610 IF X=11 THEN RESTORE 9210
2620 IF X=12 THEN RESTORE 9220
2630 IF X=13 THEN RESTORE 9230
2640 IF X=14 THEN RESTORE 9240
2650 IF X=15 THEN RESTORE 9250
2660 IF X=16 THEN RESTORE 9260
2670 IF X=17 THEN RESTORE 9270
2680 IF X=18 THEN RESTORE 9280
2690 IF X=19 THEN RESTORE 9290
2700 IF X=20 THEN RESTORE 9300
2710 RETURN
9010 DATA A,AKR,BALB/c,CBA/J,CL/Fr,C3H/He,C57BL/6,C57BL/10,C57L,DBA/1
9020 DATA DBA/2,KK,NC,NZB,NZW,P/J,RF/J,SJL,SWR/J,129
9050 DATA Idh1,Pep3,Akp1,Hc,a,Car2,Mup1,b,Gpd1,Pgm1,Ldr1,Gpi1,p
9070 DATA c,Hbb,Es1,Es2,Thy1,d,Mod1,Trf,Es3,Es10,H2K,H2D
9110 DATA a,b,b,0,a,b,a,b,b,a,a,a,+,c,d,b,b,b,+,a,b,c,a,k,d
9120 DATA b,b,b,0,a,a,a,+,b,a,a,a,+,c,d,b,b,a,+,b,b,c,b,k,k
9130 DATA a,a,b,1,+,b,a,b,b,a,a,a,+,c,d,b,b,b,+,a,b,a,a,d,d
9140 DATA b,b,a,1,+,b,a,+,b,a,a,b,+,+,d,b,b,b,+,b,a,c,b,k,k
9150 DATA a,b,b,0,at,a,a,+,b,b,a,a,+,c,d,b,b,b,+,a,b,c,.,k,d
9160 DATA a,b,b,1,+,b,a,+,b,b,a,b,+,+,d,b,b,b,+,a,b,c,b,k,k
9170 DATA a,a,a,1,a,a,b,+,a,a,a,b,+,+,s,a,b,b,+,b,b,a,a,b,b
9180 DATA a,a,a,1,a,a,b,+,a,a,a,b,+,+,s,a,b,b,+,b,b,a,a,b,b
9190 DATA b,a,a,1,a,b,b,b,a,a,a,a,+,+,s,a,b,b,+,b,b,a,a,b,b
9200 DATA b,b,a,1,a,a,a,b,a,b,a,a,+,+,d,b,b,b,d,a,b,c,b,q,q
9210 DATA b,b,a,0,a,b,a,b,b,b,a,a,+,+,d,b,b,b,d,a,b,c,b,d,d
9220 DATA a,b,b,0,a,a,b,+,a,a,b,+,c,s,b,a,b,+,a,b,c,c,b,b
9230 DATA b,b,a,0,+,a,b,b,b,a,a,a,+,+,s,b,b,b,+,a,b,c,.,d,.
9240 DATA a,c,b,0,a,a,a,+,b,b,a,a,+,+,d,b,b,b,+,b,b,c,b,d,d
9250 DATA b,b,b,1,+,a,a,+,b,b,a,a,+,c,d,b,b,b,+,a,b,c,a,z,.
9260 DATA b,c,b,1,a,a,b,b,a,b,a,a,p,+,d,b,b,a,d,b,b,a,a,p,p
9270 DATA a,b,b,0,a,a,a,+,a,a,a,a,+,c,d,b,b,a,+,a,b,b,b,b,b
9280 DATA b,b,b,1,+,b,a,+,b,b,a,a,+,c,s,b,b,b,+,a,b,c,b,s,s
9290 DATA a,b,b,0,+,b,a,+,b,b,b,b,+,c,s,b,b,b,+,a,b,c,a,q,q
9300 DATA a,b,b,1,Aw,a,a,+,a,a,a,a,p,ch,d,b,b,b,+,a,b,c,b,b,b
9990 CONSOLE C40,S0,24:END
```

References

Bailey, D. W. and B. Usama (1960). A rapid method of grafting skin on tails of mice. *Plast. Reconstr. Surg. Transplant. Bull.*, **25**: 424–425.

Carpenter, P. L. (1975). Immunology and Serology. 3rd ed. W. B. Saunders, Philadelphia.

Demant, P. (1980). Histocompatibility genes and their use in genetic control of laboratory mice. *In* A. Spiegel, S. Erichsen, and A. Solleveld (eds.) Animal Quality and Models in Biomedical Research. Gustav Fischer Verlag, Stuttgart, pp. 299–306.

Festing, M. F. W. and P. Totman (1980). Polyvalent strain-specific alloantisera as tools for routine genetic quality control of inbred and congenic strains of rats and mice. *Lab. Anim.*, **14**: 173–177.

Green, M. C. (1981). Catalog of mutant genes and polymorphic loci. *In* M. C. Green (ed.) Genetic Variants and Strains of the Laboratory Mouse. Gustav Fischer Verlag. Stuttgart, pp. 8–278.

Herzenberg, L. A. and L. A. Herzenberg (1978). Mouse immunoglobulin allotypes discription and special methodology. *In* D. W. Weir (ed.) Handbook of Experimental Immunology. Lippincott, Philadelphia, pp. 12.1–12.23.

Hoffman, H. A., K. T. Smith, J. S. Crowell, T. Nomura, and T. Tomita (1980). Genetic quality control of laboratory animals with emphasis on genetic monitoring. *In* 7th ICLAS Symp. Utrecht 1979. Gustav Fischer Verlag, Stuttgart, pp. 307–317.

Klein, J. (1975). Biology of the Mouse Histocompatibility-2 Complex, Springer-Verlag, Berlin.

Kondo, K. and K. Esaki (1962). Breeding of tester strains for coat colour genes. *Bull. Exp. Anim.*, **11**: 194–196.

Roderick, T. H., J. Staats, and J. E. Womack (1981). Strain distribution of polymorphic variants. *In* M. C. Green (ed.) Genetic Variants and Strains of the Laboratory Mouse. Gustav Fischer Verlag, Stuttgart, pp. 377–400.

Silvers, W. K. (1979). The Coat Colors of Mice. Springer Verlag, New York.

Snell, G. D. and M. Cherry (1972). Loci determining cell surface alloantigens. *In* P. Emmelot and P. Bentvelzen (eds.) RNA Viruses and Host Genome in Oncogenesis. North-Holland, Amsterdam, pp. 221–228.

III. QUANTITATIVE CHARACTERS IN GENETIC MONITORING

Quantitative characters such as skeletal measurements and breeding performance can be extremely useful in routine genetic quality control. Skeletal characters are easily measured, and with appropriate (usually multivariate) statistical analyses they can provide a highly sensitive indication of whether a group of animals of "unknown" strain are identical to samples of genetically authentic animals of that strain. Indeed, for certain types of investigation (e.g., in studying subline divergence within an inbred strain), these techniques may be more sensitive than biochemical or immunological methods because they survey simultaneously a large (though unknown) number of genetic loci. They are also of particular value for the genetic quality control of outbred stocks, which are difficult to monitor using single-gene markers, and are suitable, because they are so economical, for the routine monitoring of samples of animals from almost any type of colony. There are, however, two disadvantages to the use of such characters. First, as such traits are under the control of both genetics and environment, some "false positive" results may occur when sharp environmental effects are mistaken for genetic change. Second, the use of these characters involves a statistical interpretation of the data, which may prove to be unsatisfactory in some cases. It is therefore suggested that these highly economical quantitative characters be used for routine monitoring of colonies that have already been authenticated using biochemical and immunological methods and in which there is no special reason to suspect genetic contamination. The other more expensive techniques may then be used when problems are suspected or when the quantitative method suggests that something is wrong.

Breeding performance may also be useful, largely because it is often monitored routinely and because a cross between inbred strains usually results in "hybrid vigor," leading to an improvement in breeding performance. The use of breeding performance in routine genetic quality control is considered separately at the end of this section.

Choice of Characters

In choosing a suitable character, account should be taken of:

1. The cost of obtaining the data.
2. The amount of genetic and environmental variation associated with the chosen character(s).
3. Whether or not the animal needs to be killed.
4. Practical features such as storage of material, degree of technical expertise needed to obtain the data, and whether or not expensive equipment will be needed.
5. The extent to which the chosen characters will in fact distinguish between different strains.
6. The potential for future automation of the technique.

Fingerprints in humans fulfil most of the criteria for an ideal character. Unfortunately, rodents do not have such prints, though Festing (1974a) showed that they do have bristles on the footpads that vary in number between strains. Unfortunately, this is a univariate character (i.e., for each individual animal there is only a single numerical value) that cannot be used to distinguish among all inbred strains. Thus far no other practical external feature of the mouse or rat has been described that meets the above criteria. Internally, there are several characters such as organ weights that can be measured. For example, hematological characters are easy to measure, are multivariate, do not involve killing the animal, and are of intrinsic interest. Unfortunately, although there are undoubtedly large strain differences in these characters (Lovell *et al.*, 1981), they are far too easily influenced by environmental factors such as infection to be ideal for genetic monitoring. However, if hematological data are being collected routinely on control animals for other purposes, it might be worthwhile using discriminant function analysis, as described below, to monitor any changes in these characters. Such changes could indicate either genetic contamination or disease, which in many cases could pose an equally serious problems.

Skeletal characters are known from the work of Gruneberg (see Gruneberg, 1952) to differ markedly between inbred strains. Bones are easy to prepare and measure and, apart from the need to kill the animal, have many advantages as a source of suitable data. In a series of studies (Festing, 1972, 1973a, b, 1974b, c; Festing and Lovell, 1979), it has been shown that the shape of the right mandible of the mouse and rat, as described by 11 measurements, can, with appropriate statistical analysis, be used for routine genetic quality control. Measurements of other bones may also be used, though this is unnecessary for routine work, as the eleven mandible measurements provide sufficient information. The rest of this section is devoted to a description of the use of mandible shape for genetic monitoring, though it should again be emphasized that the same mathematical methods may be used for any multivariate set of data such as hematology or organ weight.

Laboratory Procedures

Mandible measurement

Preparation of mandibles

The mice or rats should be humanely killed, and the heads cut off. These should be boiled for a few minutes until the flesh is soft and the lower jaw can be pulled off easily without damaging the bones. Heads which have been preserved in 70% alcohol can be treated in a similar manner, though longer boiling may be needed. It is extremely difficult, however, to prepare mandibles from heads which have been preserved in formalin.

Excess flesh should be removed with forceps, and the whole lower jaw incubated with a "pinch" of powdered papain (a proteolytic enzyme) in tap water for a few hours at 30–40°C. The exact conditions are not critical, the aim being simply to obtain mandibles that are free of soft tissue. After incubation with papain, the mandibles should be clean, and the two halves of the lower jaw should have become detached. The incisor tooth should now be removed from the right mandible. At this stage the mandibles can be dried and stored prior to measurement. For long-term storage (over 2 years), however, it may be better to de-fat the mandibles in acetone for 8–10 hours and bleach them in 3% hydrogen peroxide for several hours before they are dried.

Measurement of the mandibles

Following the technique described by Festing (1979), the easiest way of measuring the mandibles is to prepare the apparatus described in Fig. 21. This consists of a photographic print of 1 mm graph paper reduced linearly to 1/4 size. This is pasted to a rigid base (plywood, plastic, or metal), and on top of it two more bits of plastic or glass are glued on as the X and Y axes. The

Quarter mm graph paper

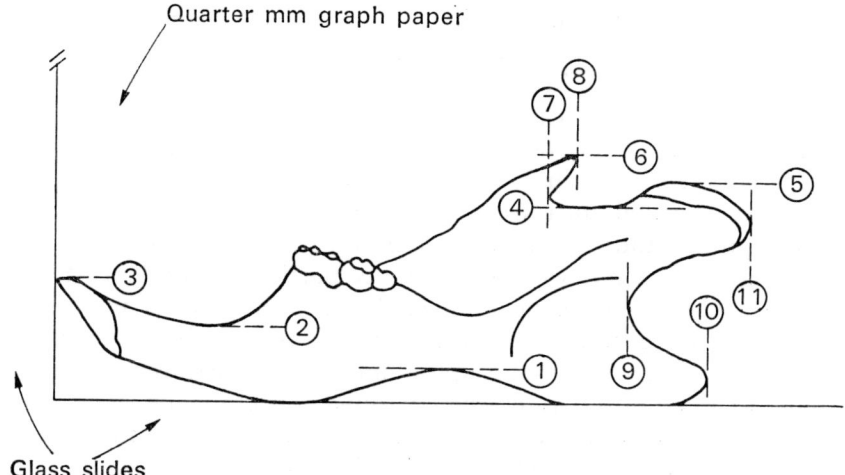

Glass slides

FIG. 21. Diagram of the 11 mandible measurements used in routine monitoring. The mandible is placed over graph paper photographically reduced to 1/4 size linearly, touching two glass or plastic axes as shown. Measurements represent the tangents of the curve of the bone relative to the graph. Measurements 1–6 represent height from the X axis, and 7–11 represent length measured from the Y axis.

surface of the print should then be covered with a glass coverslip to prevent erosion of the graph. The mandibles are placed on the apparatus, touching the X and Y axes, as shown in Fig. 21. A low power dissecting microscope (about ×10 magnification) is used to view the mandibles. The 11 measurements shown in Fig. 21 can be read off directly from the intersection of the bone with the graph. In all cases, the measurements are tangents to the curves of the bone. Measurement numbers 1–6 represent height from the X axis; numbers 7–11, length from the Y axis. The measurements are usually read to half the smallest square (i.e., to the nearest 0.125 mm), which is regarded as one unit.

The main disadvantage of this apparatus is that it is difficult to glue on the X and Y axes so that they meet exactly at right angles at the point 0,0. There is also a certain amount of measurement error in getting the bone to touch the axes exactly, and associated with parallax and subjective estimation of measurements that fall between two lines. These sources of error have not been studied in detail; however, in one unpublished study 10 mandibles from each of 5 strains chosen at random were coded and then measured twice in random order. A standard set of four discriminant functions was calculated for each. The variance was then partitioned into between-strain differences (90.2%), differences between individuals within a strain (5.3%), and differences between repeated measurements of the same mandible (4.5%). Although the study was rather small in scale, this level of measurement error does not seem to pose any particular problems. Such errors could of course be reduced by measuring each mandible more than once, but for routine work this is probably not worthwhile unless a group of mandibles "fails." The measurement of an individual mandible is extremely quick once the operator has become used to the apparatus, averaging about two minutes per mandible, one minute of which is taken up by picking up the mandible and placing it on the apparatus.

Other methods of measurement have also been developed. Hedrich (personal communication) uses a measuring microscope with crossed wires representing the same X and Y axes. The stage of the microscope is moved to position the wires in place for each measurement. When the wire is correctly aligned, the measurement is fed directly into a computer. No details have been published of the precision or speed with which this apparatus can be used. It is unlikely that it greatly increases the speed of measurement, but it probably increases precision and prevents errors in transcription of the data to the computer. The cost of this apparatus is not known, but it is probably quite high.

Festing (unpublished) has used a 35-mm photographic enlarger to cast a shadow of the mandible onto a graphics tablet attached to a microcomputer. Graph paper is aligned with the shadow, and the pen of the graphics tablet is used to transmit the data directly to the computer. Detailed studies have not yet been completed, but preliminary work shows that this method speeds up the measurement by about a factor of 2, but does not significantly increase precision.

Direct digitization of a TV scan of a mandible is already technically feasible. An example showing a scan of a C57BL/6 and a C3H/He mandible is shown in Fig. 22. Such a scan takes only 8 seconds per mandible. Programs to evaluate the scan directly have not yet been developed, but should not pose serious prob-

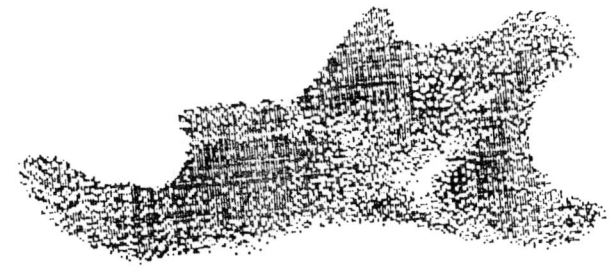

FIG. 22. Outline of a C3H/He (top) and a C57BL/6 mandible produced by a video digitizer (Microworks DS-65, PO Box 1110, Del Mar, California) from a TV picture of the mandible. Such scans take 8 seconds. The programming of a computer to provide direct identification of such pictures should not present any insuperable difficulties, though this has not yet been attempted. Note differences in shape of the coronoid (top), condyloid (top right), and angle (bottom right) processes, typical of these two strains.

lems, making it possible to automate measurement of the mandibles.

Analysis of mandible measurements

1. Correction for size: Strains differ both in the size and the shape of the mandible. However, size is probably much more subject to environmental influences than is shape, so for most routine work it is advisable to use data corrected for size. There are many ways of correcting for size. In practice, the method that has been most widely used, and which is proposed here, is to add up all eleven measurements and express each measurement as a percentage of the total. This appears to remove most of the size variation unless the groups differ markedly in size. In routine work it may also be advisable to restrict the samples to mice of, for example, 25–30 g and rats of, for example, 150–200 g in order not to get too many false positive results due to uncorrected size variation.

With certain multivariate computer programs (see below), data of this sort in which the sum of all the variables totals 100 can create problems with "colinearity." With this type of correction, there are only ten independent measures of mandible shape, as the eleventh is 100 minus the rest. In such cases, some computer programs "hang," others automatically discard one of the variables, and some are unaffected. If problems of this sort arise, it may be advisable to discard one of the variables after correcting for size but before starting the analysis.

2. *Sex:* The mandible shapes of males and females differ slightly, though this has not been studied in any great detail. For routine work, it is probably best to restrict the studies to one sex.

3. *Choice of variables:* Figure 21 gives details of the 11 measurements of the mandible described by Festing (1979). This is the maximum number of measurements that can be obtained on this bone using the techniques described here. In some cases it may be possible to get perfectly good discrimination between certain strains using only 3 or 4 measurements. However, in practice this would probably save very little time or money. The cost of the animals would be the same, as would the preparation of the mandibles and the time taken to place them on the measuring apparatus. In these circumstances, the cost of obtaining all 11 measurements is not much greater than that of obtaining a subset. Conversely, in scientific studies of subline differences or of mutagenesis, it may be worthwhile to include additional bones, but this considerably increases the cost. For routine work, the 11 measurements of the mandible appear to be adequate. However, there is nothing of great significance in these particular measurements, and any others that could be made conveniently (say as a result of automation of data collection) might be of equal value.

Strategy for setting up a routine monitoring procedure

The statistical analysis of data in order to set up a routine procedure for genetic monitoring will usually involve the following steps:

1. Ideally, the genetic authenticity of the strains to be monitored should first be checked using biochemical and immunological markers. If this is not possible, samples of about 20 animals of a specified sex and weight should be obtained from a reliable source for comparison with the colonies to be monitored.

2. Samples of about 30–50 mandibles of each strain, collected on several different occasions, and preferably from more than one source, should be measured. About two-thirds of the mandibles should be analyzed using one of the packaged discriminant function programs, some of which are mentioned below. The aim of this analysis is to obtain a set of discriminant or classification functions that will give the best possible chance of assigning "unknown" mandibles to the correct strain. The remaining third of the mandibles should then be submitted as "unknown" to see whether the calculated discriminant functions work.

3. Once a set of discriminant or classification functions has been calculated for that particular application, the coefficients can be plugged into one of the BASIC programs given in Appendix 1 (in place of the coefficients given). These programs should be easily adaptable to most microcomputers, and can be elaborated to give a direct print-out of how closely each individual matches its putative strain. It is this program that would be used for most routine work. It should require virtually no knowledge of computers to run it on almost any microcomputer (once it has been adapted to the dialect of that make of computer) and is therefore suitable for use by the technicians who are trained to measure the mandibles.

4. In some cases, it may be acceptable to use a set of discriminant functions which give good discrimination among all strains rather than just the particular strains held. Such a set is given in Table 24. However, a general set of functions of this sort will not, on average, give such good discrimination between two particular strains of interest as would a set worked out specifically for those strains. Should this strategy be adopted, then the full multivariate analysis described below would not be needed. All that would be required would be to adapt the BASIC program to one's particular computer.

Multivariate analysis

All the multivariate analyses (see below) can be carried out using commercial statistical packages such as BMDP, PSTAT, SPSS, or GENSTAT. These are available in most intermediate-to-large computer installations. The calculation of a discriminant function for just two strains can also be carried out using a computer program for multiple linear regression. Such programs are also available for many microcomputers. A discriminant function analysis using a set of previously computed discriminant functions may be carried out on a microcomputer using the BASIC programs given in the Appendix. This program can easily be altered to make use of a different set of discriminant functions based on the same 11 measurements.

The theory behind the analyses described below is given in more detail by Blackith and Reyment (1971), Afifi and Azen (1979), and Cooley and Lohnes (1971). Some of the manuals associated with the computer programs are also extremely helpful. The BMDP manual (Dixon and Brown, 1979) is particularly valuable and could usefully be consulted even if other packaged programs are actually used.

Two strains and multiple linear regression

It is easy to calculate a discriminant function for distinguishing between two inbred strains using a multiple linear regression program. The procedure is to create a new "dummy" variable that has the value 1 for one of the strains and −1 for the other or −1/n and 1/n if sample size differ. This is then regarded as the dependent variable in the analysis. The program then attempts to use the mandible measurements to predict group membership. This may be understood more clearly by reference to an example.

Appendix 2 lists the 11 measurements of 100 mouse mandibles. These comprise samples of 10 mice from 5 different strains sampled on two occasions during 1981. These raw data are provided for anyone who wishes to repeat some of the analyses given here. For this example, sample numbers 1839 (strain A/Lac) and 1840 (SWR) were chosen for analysis using the PSTAT computer package (Buhler and Buhler, 1979). The data were first corrected for size by expressing each as a percent of the sum, as described. Variable number X11 was then dropped from the analysis to avoid colinearity. The size-corrected data are shown in Table 17. In-

TABLE 17. Mandible measurements of samples 1839 and 1840 corrected for size

Strain	N	X1	X2	X3	X4	X5	X6	X7	X8	X9	X10	X11
A. Lac	1	0.94	2.64	3.77	6.59	7.53	8.29	11.30	12.05	14.31	16.38	16.20
A. Lac	2	1.13	2.82	3.77	6.78	7.53	8.29	11.11	11.86	14.12	16.38	16.20
A. Lac	3	1.12	2.79	4.09	6.51	7.43	8.18	11.34	12.27	13.94	16.36	15.99
A. Lac	4	1.06	2.66	3.72	6.38	7.27	8.16	10.99	11.88	14.18	16.31	17.38
A. Lac	5	1.13	2.84	3.97	6.43	7.18	8.13	11.34	12.10	14.37	16.26	16.26
A. Lac	6	0.95	2.66	3.80	6.45	7.21	8.16	11.39	12.33	14.42	16.70	15.94
A. Lac	7	1.11	2.60	3.71	6.49	7.24	7.98	11.50	12.24	14.29	16.70	16.14
A. Lac	8	1.09	2.74	4.01	6.57	7.48	8.39	11.13	12.23	14.23	16.42	15.69
A. Lac	9	1.34	2.68	3.82	6.50	7.27	8.41	11.47	12.24	14.15	16.06	16.06
A. Lac	10	1.08	2.71	3.97	6.50	7.58	8.30	11.37	12.09	14.26	16.25	15.88
SWR	1	0.73	2.90	4.36	5.99	7.08	7.62	11.98	12.52	14.16	15.97	16.70
SWR	2	0.75	2.81	4.12	5.99	6.74	7.49	11.99	12.55	14.42	16.10	17.04
SWR	3	0.73	2.73	4.01	6.19	6.92	7.65	12.02	12.75	14.21	16.03	16.76
SWR	4	0.76	2.66	3.99	6.08	6.84	7.60	11.79	12.36	14.45	16.35	17.11
SWR	5	0.70	2.82	4.23	5.81	6.87	7.39	12.15	12.68	14.61	15.85	16.90
SWR	6	0.72	2.89	4.34	5.97	6.87	7.59	11.93	12.48	14.29	16.09	16.82
SWR	7	0.73	2.73	4.00	6.18	7.09	7.82	12.00	12.55	14.18	16.00	16.73
SWR	8	0.73	2.91	4.18	6.18	6.91	7.64	12.00	12.55	14.18	16.00	16.73
SWR	9	0.72	2.70	4.14	6.12	7.01	7.55	12.05	12.23	14.21	16.19	17.09
SWR	10	0.71	2.85	3.92	6.24	6.95	7.66	12.12	12.48	14.26	16.04	16.76

TABLE 18. Output from PSTAT multiple linear regression program for discriminating between two groups

Final Summary of Regression on Dependent Variable   SN

Multiple R	0.9898
Mult R squared	0.9797
Std. error of est.	0.2124
Constant	31.4152

Analysis of variance	DF	Sum of squares	Mean square	F ratio	Prob. level
Regression	10.	19.594	1.959	43.430	0.000
Residual	9.	0.406	0.045		

Step	Num Var Now in	Mult R	Mult R sq.	Change in R sq	Variable entered (*shows deleted)	B, raw coefficient	Stand error of B
1	1	0.927	0.859	0.859	X1	−2.4116	0.6068
2	2	0.934	0.873	0.013	X2	−1.4873	1.1410
3	3	0.939	0.883	0.010	X3	1.0833	0.8066
4	4	0.955	0.911	0.029	X4	1.2251	1.1048
5	5	0.966	0.933	0.022	X5	−1.5843	0.6484
6	6	0.967	0.935	0.002	X6	−0.3820	0.7604
7	7	0.978	0.956	0.021	X7	0.5186	0.5003
8	8	0.980	0.960	0.004	X8	−0.3872	0.4525
9	9	0.985	0.969	0.009	X9	−0.5020	0.4796
10	10	0.990	0.980	0.010	X10	−1.0428	0.4890
					constant	31.41518	

Step	Variable entered (* shows deleted)	Beta, stand. coefficient	Final F to delete	F when entered or deleted	Simple cor. with dep.	Partial cor. in final step
1	X1	−0.4793	15.796	109.923	−0.9270	−0.7981
2	X2	−0.1384	1.699	1.787	0.4743	−0.3985
3	X3	0.2051	1.804	1.358	0.6997	0.4086
4	X4	0.3074	1.230	4.838	−0.8863	0.3467
5	X5	−0.4047	5.970	4.589	−0.8689	−0.6315
6	X6	−0.1275	0.252	0.330	−0.9370	−0.1652
7	X7	0.1954	1.074	5.871	0.9393	0.3265
8	X8	−0.0934	0.732	1.179	0.7944	−0.2743
9	X9	−0.0705	1.096	2.935	0.2438	−0.3294
10	X10	−0.2356	4.547	4.547	−0.7075	−0.5794

spection of these data shows large strain differences with no overlap for several of the variables (e.g., numbers 1, 4, and 7). The program provides a step-wise method for sorting through the 10 variables to see which is most important for discriminating between strains. This could be used to find a subset of the variables that would distinguish between the strains, thereby reducing the number of measurements needed per mandible. However, for reasons already stated, all measurements were retained, and the stepping procedure was not used.

Table 18 shows the Final Summary of Regression Analysis on Dependent Variable SN. The dummy variable SN was given the value of $+1$ for strain SWR and -1 for A/Lac. Space does not allow a full description of the entire output. Much of it is concerned with the relative importance of each of the variables. For our purposes, the most important part of the output is the column headed "B, Raw Coefficient," as these are the coefficients for the discriminant function that gives maximum discrimination between the two strains and that can be used in the future to assign "unknown" mandibles to one of the two groups. The unknown mandible is given a score (called D) calculated as follows:

$$D = -2.4116 \times X1 - 1.4873 \times X2 + 1.0833 \times X3 \ldots - 1.0428 \times X10 + 31.41518$$

where X1, X2 . . . X10 are the first ten measurements corrected for size, and the numbers are the "B, Raw Coefficients."

If D should turn out to be positive for an unknown mandible, then it would be assigned to strain SWR, and, if negative, to strain A/Lac. Sample numbers 1768 and 1780 were then submitted as "unknown" to test the calculated discriminant functions. The results are shown in Table 19, in the column headed "Pre SN" (for "predicted strain"). It can be seen that all the SWR had positive values and so were correctly assigned to strain SWR. Similarly, 8/10 of the strain A mice had negative values and were correctly assigned to that strain. However, two had positive values and thus would be assigned to SWR and therefore misclassified. Assuming all mice are in fact genuine, this gives an overall classification of 18/20 or 90% correct. If one of these colonies became genetically contaminated with mice from the other colony, mandible shape of the contaminated colony would probably shift towards the mean of the two colonies, and the percentage of incorrect classifications would increase, indicating that a genetic change might have occurred.

Finally, in this example the discriminant function was set up using only ten mandibles of each strain, sampled on a single occasion. In practice, much larger sample sizes collected on more

TABLE 19. Predicted group membership on a sample of "unknown" mice of strains SWR and A

SN	X1	X2	X3	X4	X5	X6	X7	X8	X9	X10	X11	Pre SN*
SWR	0.78	2.91	4.26	6.20	6.78	7.75	11.82	12.40	14.34	16.09	16.67	1.09
SWR	0.76	2.86	4.19	6.10	6.48	7.62	11.81	12.57	14.48	16.19	16.95	1.27
SWR	0.77	2.87	4.21	6.13	6.70	7.66	11.88	12.64	14.18	16.09	16.86	1.18
SWR	0.78	2.71	4.07	6.20	6.98	7.75	11.63	12.40	14.53	16.28	16.67	0.46
SWR	0.76	2.85	4.17	6.26	6.83	7.59	11.95	12.52	14.42	15.94	16.70	1.31
SWR	0.76	2.86	4.38	6.10	6.86	7.62	11.81	12.19	14.29	16.19	16.95	1.12
SWR	0.77	2.89	4.24	6.36	6.94	7.71	11.95	12.33	14.26	15.99	16.57	1.29
SWR	0.75	2.82	4.14	6.21	6.97	7.72	11.68	12.43	14.31	16.20	16.76	0.68
SWR	0.75	2.64	3.77	6.23	6.98	7.74	11.89	12.45	14.34	16.23	16.98	0.59
SWR	0.76	2.65	3.97	6.43	6.99	7.75	11.72	12.48	14.37	16.26	16.64	0.87
A	0.80	2.79	3.99	6.39	7.19	7.98	11.58	12.18	14.37	16.37	16.37	0.06
A	0.97	2.71	3.88	6.40	7.17	8.14	11.43	12.40	14.15	16.47	16.28	−0.55
A	1.15	2.69	3.65	6.72	7.49	8.45	11.32	11.90	14.20	16.51	15.93	−1.34
A	0.81	2.82	4.03	6.45	6.85	7.86	11.49	12.10	14.31	16.94	16.33	0.11
A	0.80	2.61	3.81	6.61	7.82	8.42	11.02	12.02	14.23	16.63	16.03	−1.19
A	1.20	2.79	3.78	6.57	7.37	8.37	11.35	12.35	13.94	16.33	15.94	−1.26
A	1.15	2.68	3.64	6.70	7.66	8.43	11.30	11.88	13.98	16.48	16.09	−1.49
A	0.97	2.73	3.90	6.43	7.41	8.19	11.50	11.89	14.42	16.37	16.18	−0.71
A	1.13	2.64	3.77	6.60	7.36	8.30	11.32	11.89	14.15	16.60	16.23	−1.05
A	0.97	2.71	3.68	6.58	7.54	8.12	11.41	11.99	14.31	16.63	16.05	−1.22

* Positive values, strain SWR predicted; negative values strain A/Lac predicted.

than one occasion should be used, and this should lead to much better discrimination.

Multiple discriminant functions involving three or more strains

If three or more strains are involved, it is necessary to use a specialized multiple discriminant function program rather than one for multiple linear regression. Strictly speaking, with more than two groups the technique is known as "canonical variate" rather than discriminant function analysis. With N groups it is possible to calculate N-1 discriminant functions. The first gives the best possible discrimination between the groups, with the second giving the best discrimination using the remaining information. With several groups, the first few functions will normally provide most of the discrimination. The program must also have some method of assigning unknown individuals to a group in the same way as with the two-group example given above in which all animals with a positive discriminant score were assigned to one group and those with a negative score to the other. The usual method is to calculate a "classification function" which assigns a score for the closeness of each individual to each group. The individual is eventually assigned to that group for which it has the highest score.

An example of an analysis using the PSTAT package is presented in Table 20. In this analysis all the data on the 100 mandibles (corrected for size) given in the appendix were used to set up the discriminant functions, and data on another 80 mandibles (data not given) were used to validate the results against "unknown" mandibles. The output of the program has been reduced to save space. Thus the program gives means for each variable for each group, which are not shown in the table. The univariate and multivariate F values provide an analysis of the differences between strain means for each variable. These are not of much importance in this case, because inspection of the data is sufficient to show that strains differ. The PSTAT program has a stepping procedure for identifying those variables which are of least value in distinguishing among strains, and eliminating them from the analysis. For reasons already stated, this option was not used and all 10 variables were retained. Again, variable X11 was dropped after correcting for size to avoid colinearity problems.

The classification table gives the best visual impression of the success of the analysis. For example, out of the 20 B10 mandibles, 19 had a higher score when compared with the B10 group than any other group, while one scored higher with strain B6. In this table,

TABLE 20. Multiple discriminant analysis. Abbreviated output
of the PSTAT program

Multiple discriminant analysis of mandible measurements using data corrected for size. Variable X11 dropped due to colinearity. Two independent samples each of 10 mandibles for each of the five strains were used.

Means of each variable for each strain not given in order to save space.

		Univ. F	Univ. prob.	Mult. F	Mult. prob.
1	X1	13.97	0.0000	6.42	0.0001
2	X2	23.38	0.0000	3.94	0.0055
3	X3 W	4.01	0.0048	2.04	0.0958
4	X4	3.00	0.0222	7.25	0.0000
5	X5	6.40	0.0001	10.70	0.0000
6	X6	2.45	0.0510	5.67	0.0004
7	X7	3.91	0.0055	2.62	0.0404
8	X8	21.92	0.0000	2.99	0.0232
9	X9	50.61	0.0000	6.90	0.0001
10	X10	48.29	0.0000	45.63	0.0000

Univariate degrees of freedom are 4 and 95.
Multivariate degrees of freedom are 4 and 95.
Wilks lambda=0.01804; approx. F=15.44; prob=0.000.
DF=40 and 328.0.
Ending step 1 with 10 variables in use.
Stepping completed.

Generalized mahalanobis D square=875.33 with 40 degree of freedom.
Interpreted as a chi square. The D square values indicates a probability of 0.0000 that the mean values are the same in all the group for each of the variables.

Classification table

Original Group		Assigned groups based on largest score				
		1	2	3	4	5
		B10	BALB	B6	A	SWR
1	B10	19	0	1	0	0
2	BALB	0	20	0	0	0
3	B6	2	0	18	0	0
4	A	0	0	0	19	1
5	SWR	0	1	1	0	18

94 out of 100 were assigned to their original group, that is, 94.0%.

TABLE 21. Predicted group membership of "unknown" mice of five
strains using the PSTAT program

Classification table

Original	Group	\multicolumn{5}{c}{Assigned groups based on largest score}
		1	2	3	4	5
		B10	BALB	B6	A	SWR
1	B10	14	0	6	0	0
2	BALB	0	20	0	0	0
3	B6	2	0	16	0	2
4	A	0	0	0	8	2
5	SWR	0	0	0	0	10

68 out of 80 were assigned to their original group, that is, 85.0%.

TABLE 22. F-Matrix for comparing group centroids
from program BMDP7M

	B6	SWR	A	B10	BALB	B10T	BALBT	B6T	AT
SWR	15.47								
A	15.20	12.71							
B10	4.69	15.01	25.55						
BALB	35.80	18.44	31.05	24.30					
B10T	4.26	14.81	25.26	0.38	26.07				
BALBT	38.02	19.09	35.46	24.41	1.83	25.20			
B6T	1.27	10.67	14.65	4.69	31.29	3.48	31.94		
AT	14.93	7.09	3.20	20.49	20.23	19.67	21.07	12.6	
SWRT	13.81	2.02	11.40	12.35	13.82	12.13	13.28	10.6	6.67

Note: Critical value for F at the 5% level is 1.92.
Degree of freedom = 10 and 86.

TABLE 23. Calculated discriminant (or, more strictly, classification)
functions from the PSTAT program

	B10	BALB	B6	A	SWR
X1	−2.9471	−20.38814	6.3293	17.2890	−0.2831
X2	8.2482	−13.28343	11.8212	8.4271	−15.2131
X3	−10.3067	0.59531	−1.3926	10.3449	0.7591
X4	−15.5871	7.45724	−12.8061	7.4512	13.4847
X5	2.5816	−0.40224	5.0327	7.1986	−14.4107
X6	−3.3570	0.80001	−0.0080	6.6936	−4.1285
X7	−6.3113	1.10441	−4.3946	10.6868	−1.0853
X8	−0.0955	−12.12116	5.5080	6.9858	−0.2771
X9	−5.6328	19.24441	−14.2229	1.3737	−0.7624
X10	−14.8352	−10.52022	1.0013	26.4625	−2.1084
Constant	522.8022	28.55451	183.1006	−888.8552	154.3979

individuals falling on the main diagonal were correctly classified, while those off the diagonal were incorrectly classified. Note that the analysis could distinguish reasonably well (i.e., in 37/40 cases) between the B10 and B6 sublines. Overall, 95% of the mandibles were correctly classified. However, the success of the discriminant analysis should really be judged by the ability of the calculated functions to classify "unknown" mandibles which were not used in setting up the original groups. The results of such a further analysis are given in Table 21, in which 80 mandibles of these five strains were submitted as "unknown." In this case, 68/80 (85%) were correctly classified. Of the 12 mandibles that were misclassified, 8 represented misclassification between the closely related B6 and B10 sublines, 2 were as a result of B6 mandibles being classified as SWR, and 2 as a result of strain A mandibles being mistaken for SWR. It is probable that better discrimination could have been achieved with larger samples of mice, but these data probably provide an adequate basis for routine work using the BASIC program described below. The BMDP7M computer program produces exactly the same classification matrix with these data, as well as a great deal of other output. A table of F-values comparing the centroids of each group is of particular interest as it shows approximately how similar the mandible shape of each strain pair is (Table 22). (Note that "T" indicates test group not involved in calculating the DFs). It is immediately clear that strains B6 and B10 are very similar, even though they can be distinguished. The samples of SWR and A in the "unknown" test group also differed significantly from the original sample. This is a reflection of the small sample sizes and possible environmental influences. The PSTAT program also produces a file of "calculated discriminant functions," as given in Table 23. These are analogous to the "classification functions" described by Dixon and Brown (1979) for the BMDP7M program. These functions can be used to classify future groups either by using the original package programs or by developing a program specifically for routine work. The use of such a program is discussed below.

Another option is to use the discriminant functions given in the BMDP7M program output as "coefficients for canonical variables." These can be used in a routine program such as the second BASIC program given in the Appendix to reduce the dimensionality of the data so that the positions of each strain relative to

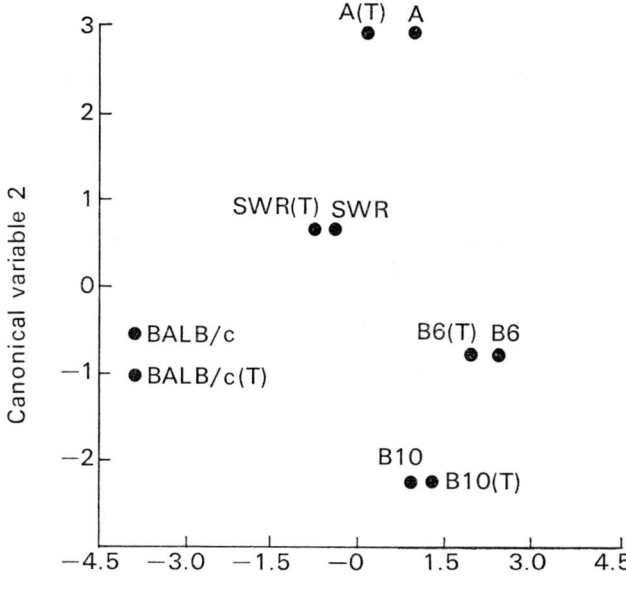

FIG. 23. Graph of first versus second canonical variable (or discriminant function) for the five strains given in the example, and for the five test groups (T) not used in the calculations. Each point is the mean of 10–20 mandibles. Note that the test samples are in close agreement with the original groups, that B6 and B10 are close but different, and that SWR has the most "average" mandibles.

each other are clear. This is demonstrated in Fig. 23, which shows plots of the first two discriminant functions for each of the five strains and each of the five "test samples" of this example. This figure shows clearly that repeat samples of the same strain are very close to each other, that B6 and B10 are close, and that in this case strain SWR is approximately in the middle of the five groups, so that some misclassification is likely to involve that strain. The second BASIC program has been written with the actual coefficients obtained in this example, though it is recommended that these coefficients be re-calculated for each particular application. The first four discriminant functions (assuming 5 or more strains) will usually provide sufficient information for most purposes. Also incorporated in the program is a procedure for testing how close a sample of mandibles is to "background" or previously collected data. This is simply the sum of squares of the differences between the discriminant scores of the "background data" and the new sample. This closely approximates a chi-squared distribution, and with sample sizes of 10 individuals it has been shown empirically (Festing and Lovell, 1979) that with "genuine" samples the value of 12 will be exceeded only about 3% of the time. However, those who use this method (which on balance may be better than the use of the classification function) will gradually gain experience over a period of time in knowing how large the "chi-squared" value has to be to indicate that further action is necessary.

Two BASIC programs are given in Appendix 1. These are interactive programs designed to be used by people with no experience in the use of computers. They have been written as simply as possible with numerous REM statements, and are largely self-explanatory. The two programs are also very similar, though the output is different. Both programs may easily be elaborated. For example, printed output is usually required, and a programmer should have no difficulty in adding suitable printing instructions.

1. The classification program: Statements 130–170 contain the coefficients for the classification functions from the PSTAT program given in Table 23. These and statement 180 could be altered to obtain optimum discrimination for any particular application, following an analysis using PSTAT or BMDP7M.

The input consists of the raw mandible measurements (not size-corrected) for a new test sample (up to 30 mandibles, though 10 would be usual). The output shows how many of the test mandibles have the highest score for each of the "background data" strains. The user will have to decide what degree of misclassification is acceptable, as this will depend on the extent to which the strains involved actually differ.

2. The discriminant function program: In this program, statements 130–160 contain constants obtained by submitting the data in the Appendix to BMDP7M. These were listed as the Coefficients for Canonical Variables. Input data are again the raw mandible measurements. In this particular case, all 11 measurements were given to the BMDP program, but X3 was rejected due to colinearity; hence coefficients associated with X3 in the program are 0.0. The output is the mean of the four discriminant functions for the group. It would be easy to add a standard deviation. The program also has provision for adding the (in this case) four DFs from previous runs (these could be accumulated for each strain), giving the chi-square statistic. Again, the user will have to obtain experience of how good this fit must be; but as a starting point, if it is less than about 12 units, the fit will usually be acceptable.

A further option

Another option, discussed in detail by Festing (1979), is to use a set of canonical variables previously calculated for a representative set of inbred strains. This was done using BMDP7M program with 20 different inbred strains. The first four "coefficients for canonical variables" can now be used to reduce the eleven mandi-

TABLE 24.   Coefficients for calculating four discriminant functions, based on a representative sample of strains of mice and rats

Mice

Variable	DF1	DF2	DF3	DF4
1	3.62	9.00	5.04	−0.60
2	5.76	5.79	−3.21	4.92
3	2.14	1.18	−3.96	−6.51
4	0.14	−1.18	2.28	0.63
5	3.30	−1.52	0.66	−0.69
6	−0.40	2.00	5.38	−8.90
7	−3.23	−0.25	0.20	−2.04
8	6.40	0.31	4.93	1.55
9	−5.99	3.30	1.23	2.27
10	2.73	6.52	1.49	−5.16
11	5.34	2.73	4.15	−0.56
Constant	−136.66	−226.83	−216.11	151.61

Rats

Variable	DF1	DF2	DF3	DF4
1	5.84	−11.60	−0.77	4.15
2	1.51	−11.81	16.37	−7.81
3	4.75	9.30	−5.03	−0.30
4	1.64	0.30	−1.44	6.80
5	−7.90	−3.96	−3.51	−4.93
6	−4.35	−5.78	3.16	−11.15
7	−7.38	−5.95	−0.30	−2.77
8	−6.42	−1.83	2.82	0.49
9	0.00	0.00	0.00	0.00
10	−1.45	−1.45	−0.81	−5.29
11	−1.71	−3.60	6.46	−4.20
Constant	273.80	252.90	−139.53	277.83

ble measurements of any strain to four "discriminant functions" for each individual. The coefficients calculated by Festing (1979) and a similar set for rats (unpublished data) are given in Table 24. All that is necessary is to plug them into statements 130–160 of program 2 in place of the constants given. No multivariate analysis is necessary. However, the use of a representative set of coefficients will not be as good in most cases as a set tailored specifically to the strains to be monitored.

Breeding performance

Breeding performance is a character which should be monitored as part of good routine husbandry and management practices. Although it cannot be relied on as the sole means of genetic monitoring, it can be extremely useful in some circumstances, as outlined below.

Inbred strains show relatively poor breeding performance as a result of "inbreeding depression" arising from having a number of mildly deleterious recessive genes in a homozygous state. F_1 hybrids (the first-generation cross between two inbred strains), however, show excellent breeding performance due to "hybrid vigor," the opposite of inbreeding depression. Table 25 shows the results of a study of breeding performance in 5 inbred strains

TABLE 25. Genetic monitoring by litter size

	1st litter size
5 Inbred strains	6.1 ± 2.00
20 F_1 hybrids	9.5 ± 2.81
Heterosis	3.4 young

Source: Festing, 1976.

TABLE 26. Distribution of litter size in inbreds and F_1 hybrids

No. young over mean (x)	Percent over $(\text{mean} + x)$	
	Inbreds	F_1 Hybrids
0	50	89
1	38	80
2	16	70
3	7	66
4	2	42
5	0.6	29
6	0.1	18

and the 20 F_1 hybrids produced by crossing the strains in all possible combinations (Festing, 1976). Hybrid vigor resulted in an average of 3.4 extra young in the first litters, an increase of about 50%.

In Table 26, it has been assumed that the first litter size has been characterized for each breeding colony. As litter size has an approximately normal distribution, about 50% of inbreds will fall on each side of the mean. On the other hand, the table shows that 89% of any F_1 hybrids would have a litter size larger than the mean of the inbreds. More important, less than 1% of the inbreds would have a litter size of the mean plus 5 pups, whereas 29% of the hybrids would exceed this value (these figures assume that the hybrids behave in the same way as those studied by Festing [1976]).

Of course, if an individual has a large litter size, this does not mean that it is the result of genetic contamination. On the other hand, if the average breeding performance of a small colony increases substantially, and this coincides with the establishment of new matings, then these new matings should be regarded with suspicion.

Breeding performance could be used most effectively in routine monitoring as a method of picking out those offspring that are most likely to be the result of genetic contamination. For example, if any litter should exceed the strain average by, say, 5 pups, then one or more of the offspring should be checked for biochemical or immunological markers. Part of the animal house routine could be to mark the cage of any such female and place her offspring in a separate box until they have been checked. Such a procedure should substantially increase the efficiency of the routine monitoring methods outlined elsewhere in this book.

Appendix 1

Two basic programs designed for routine interractive use on an Apple II® microcomputer, but adaptable to most types of micro-computers. These programs should be easy to expand to give printed output.

```
1 REM DISCRIMINANT FUNCTION PROGRAM
5 HOME:CLEAR
10 DIM X(10,12),S$(10),Y(30,12),SC(10,31),CT(10)
20 PRINT "ROUTINE GENETIC MONITORING"
30 PRINT "USING 11 MANDIBLE MEASUREMENTS"
40 PRINT:PRINT "WRITTEN BY MICHAEL FESTING"
50 PRINT "ON AN APPLE 2 COMPUTER"
100 REM NF IS NUMBER OF FUNCTIONS (MAX 10) *********
110 NF=4
120 REM THE CONSTANTS GO HEAR *********
130 DATA 4.998,4.7984,0,-3.025,1.5935,0.353,-0.199,3.206,-4.701,2.638,0.279,-29
857
140 DATA 0.441,-4.254,0,1.319,-3.67,-2.079,-0.506,-2.142,-2.429,3.765,-3.564,11
1.341
150 DATA -0.65,4.927,0,-5.186,5.597,1.417,0.633,-1.279,0.695,2.381,-1.312,-50.5
03
160 DATA -10.097,-10.497,0,-2.502,-7.104,-4.681,-8.357,-5.917,-9.649,-4.81,-4.6
99,604.732
190 REM READ DATA INTO ARRAY X(I,J) *********
200 FOR I=1 TO NF
210 FOR J=1 TO 12
220 READ X(I,J)
230 NEXT J:NEXT I
300 REM DATA ENTRY STARTS HERE *********
310 PRINT:INPUT "SAMPLE OF STRAIN ";SN$
320 PRINT:INPUT "NUMBER IN SAMPLE ";GN
330 FOR I=1 TO GN
340 FOR J=1 TO 11
350 PRINT "ANIMAL NO    ";I
360 PRINT "MEASUREMENT NO-";J
370 INPUT "               MEASUREMENT=";Y(I,J)
380 NEXT J:PRINT
400 REM INPUT CORRECTIONS *********
420 PRINT "ANIMAL NO=";I
430 FOR J=1 TO 11
440 PRINT J,Y(I,J)
450 NEXT J
460 PRINT "PRESS 0 FOR NO CORRECTIONS"
470 PRINT "ELSE MEASRT. NO. FOR CORRN"
480 INPUT Z
490 IF Z=0 THEN 550
500 PRINT "MEASUREMENT NO   ";Z
510 INPUT "CORRECT VALUE  =";W
520 Y(I,Z)=W
530 GOTO 420
550 NEXT I
```

```
600 REM CORRECT FOR SIZE *********
610 FOR I=1 TO GN
620 FOR J=1 TO 11
630 Y(I,12)=Y(I,12)+Y(I,J)
640 NEXT J
650 FOR J=1 TO 11
660 Y(I,J)=100*Y(I,J)/Y(I,12)
670 NEXT J:NEXT I
680 PRINT "CALCULATING"
700 REM CALCULATE ARRAY OF SCORES *********
710 FOR I=1 TO GN
715 Y(I,11)=1
720 FOR K=1 TO GN
730 FOR J=1 TO 11
740 SC(K,I)=SC(K,I)+Y(I,J)*X(K,J)
750 NEXT J:NEXT K:NEXT I
800 REM MEAN CSORES AND CLASSIFICATION *********
810 FOR I=1 TO GN
820 T1=SC(1,I)
830 T2=1
840 FOR K=1 TO NF
850 IF SC(K,I)<T1 THEN 870
860 T1=SC(K,I):T2=K
870 SC(K,31)=SC(K,31)+SC(K,I)
900 NEXT K
910 CT(T2)=CT(T2)+1
920 NEXT I
1000 REM OUTPUT *********
1010 HOME:PRINT "PAMPLE OF ";GN;" MICE OF STRAIN ";SN$
1015 PRINT:PRINT
1020 PRINT "STRAIN    NO ASSIGNED      MEAN SCORE"
1030 FOR I=1 TO NF
1040 A=INT(SC(I,31)*100/GN):A=A/100
1050 PRINT S$(I),CT(I),A
1060 NEXT I
1070 INPUT "ANOTHER GROUP ? Y/N";X$
1080 IF X$="N" THEN 1100
1090 IF X$="Y" THEN 5
1100 END
```

```
1 REM CLASSIFICATION PROGRAM
5 HOME:CLEAR
10 DIM X(10,11),S$(10),Y(30,12),SC(10,31),CT(10)
20 PRINT "ROUTINE GENETIC MONITORING"
30 PRINT "USING 11 MANDIBLE MEASUREMENTS"
40 PRINT:PRINT "WRITTEN BY MICHAEL FESTING"
50 PRINT "ON AN APPLE 2 COMPUTER"
100 REM NF IS NUMBER OF FUNCTIONS (MAX 10) *********
110 NF=5
120 REM THE CONSTANTS GO HERE FOLLOWED BY RELEVENT STRAIN NAME *********
130 DATA -2.9471,8.2482,-10.3067,-15.5871,2.5816,-3.357,-6.3113,-0.0955,-5.6328
,-14.832,522.8022
140 DATA -20.38814,-13.28343,0.59531,7.45724,-0.40224,0.80001,1.10441,-12.12116
,19.24441,-10.52022,28.55451
150 DATA 6.3293,11.8212,-1.3926,-12.8061,5.0327,-0.008,-4.3946,5.508,-14.2229,1
.0013,183.1006
160 DATA 17.2890,8.4271,10.3449,7.4512,7.1986,6.6936,10.6868,6.9858,1.3737,26.4
625,-888.8552
170 DATA -0.2831,-15.2131,0.7591,13.4847,-14.4107,-4.1285,-1.0853,-0.2771,-0.76
24,-2.1084,154.3979
180 DATA B10,BALB,B6,A,SWR
190 REM READ DATA INTO ARRAY X(I,J) *********
200 FOR I=1 TO NF
210 FOR J=1 TO 11
220 READ X(I,J)
230 NEXT J:NEXT I
240 FOR I=1 TO NF
250 READ S$(I)
260 NEXT I
300 REM DATA ENTRY STARTS HERE *********
310 PRINT:INPUT "SAMPLE OF STRAIN ";SN$
320 PRINT:INPUT "NUMBER IN SAMPLE ";GN
330 FOR I=1 TO GN
340 FOR J=1 TO 11
350 PRINT "ANIMAL NO    ";I
360 PRINT "MEASUREMENT NO-";J
370 INPUT "          MEASURMENT=";Y(I,J)
380 NEXT J:PRINT
400 REM INPUT CORRECTIONS *********
420 PRINT "ANIMAL NO=";I
430 FOR J=1 TO 11
440 PRINT J,Y(I,J)
450 NEXT J
460 PRINT "PRESS 0 FOR NO CORRECTIONS"
470 PRINT "ELSE MEASRT. NO. FOR CORRN"
480 INPUT Z
490 IF Z=0 THEN 550
500 PRINT "MEASUREMENT NO   ";Z
510 INPUT "CORRECT VALUE   =";W
520 Y(I,Z)=W
530 GOTO 420
550 NEXT I
```

```
600 REM    CORRECT FOR SIZE *********
610 FOR I=1 TO GN
620 FOR J=1 TO 11
630 Y(1,12)=Y(I,12)+Y(I,J).
640 NEXT J
650 FOR J=1 TO 11
660 Y(I,J)=100*Y(I,J)/Y(I,12)
670 NEXT J:NEXT I
680 PRINT "CALCULATING"
700 REM    CALCULATE ARRAY OF DISCRIMINANT SCORES *********
710 FOR I=1 TO GN
715 Y(I,12)=1
720 FOR K=1 TO NF
730 FOR J=1 TO 12
740 SC(K,I)=SC(K,I)+Y(I,J)*X(K,J)
750 NEXT J:NEXT K:NEXT I
800 REM    MEAN SCORES *********
810 FOR I=1 TO GN
840 FOR K=1 TO NF
870 SC(K,31)=SC(K,31)+SC(K,I)
900 NEXT K
920 NEXT I
1000 REM    OUTPUT *********
1010 HOME:PRINT "SAMPLE OF ";GN;" MICE OF STRAIN ";SN$
1015 PRINT:PRINT
1017 REM    DF OF BACKGROUND IS ZERO AT THIS STAGE
1020 PRINT "DF NO      MEAN   DF    DF OF BACKGRND"
1030 FOR I=1 TO NF
1040 A=INT(SC(I,31)*100/GN):A=A/100
1050 PRINT I,A,CT(I)
1060 NEXT I
1065 PRINT "CHI-SQ. STAT. FOR STRAIN ";S$;"=";CH
1070 PRINT:INPUT "ENTER E TO END OR C TO CALCULATE CHI-SQ ";X$
1080 IF X$="E" THEN 2000
1090 IF X$="C" THEN 1110
1100 GOTO 1070
1105 REM    CALCULATION OF CHI-SQ. FOR COMPARISON WITH PREVIOUS DATA ********
1110 INPUT "STRAIN NAME ";S$:CH=0
1120 FOR I=1 TO NF
1130 PRINT "DF NUMBER ";I
1140 INPUT CT(I)
1150 NEXT I
1160 FOR I=1 TO NF
1170 CH=CH+(CT(I)-SC(I,31)/GN)^2
1180 NEXT I
1190 GOTO 1010
1200 GOTO 1070
2000 END
```

Appendix 2

Raw data on measurements of 100 mandibles used in the examples

Sample number	Strain	Date collected
1767	C57BL/6	51/81
1768	SWR	51/81
1780	A/Lac	51/81
1781	C57BL/10	51/81
1782	BALB/c	51/81
1835	C57BL/10	138/81
1836	BALB/c	138/81
1838	SWR	138/81
1839	A/Lac	138/81
1841	C57BL/6	138/81

All mice came from the same commercial breeder and are believed to be genuine samples of the stated strains.

Raw mandible measurements used in the calculations
SN=sample number, N=Individual

SN	N	X1	X2	X3	X4	X5	X6	X7	X8	X9	X10	X11
1767	1	6	16	22	34	39	44	62	66	76	86	89
1767	2	6	16	22	34	39	44	62	66	76	87	91
1767	3	6	16	22	35	40	45	62	66	76	86	89
1767	4	4	15	21	34	39	44	63	66	76	87	90
1767	5	6	16	22	35	40	44	62	68	76	86	88
1767	6	6	16	22	34	38	44	62	66	76	87	89
1767	7	5	16	22	35	40	44	60	66	74	84	89
1767	8	6	17	23	36	40	46	61	66	75	86	88
1767	9	5	16	22	33	39	44	61	66	76	86	88
1767	10	5	16	22	33	37	43	61	67	75	86	88
1768	1	4	15	22	32	35	40	61	64	74	83	86
1768	2	4	15	22	32	34	40	62	66	76	85	89
1768	3	4	15	22	32	35	40	62	66	74	84	88
1768	4	4	14	21	32	36	40	60	64	75	84	86

Raw mandible data (continued)

SN	N	X1	X2	X3	X4	X5	X6	X7	X8	X9	X10	X11
1768	5	4	15	22	33	36	40	63	66	76	84	88
1768	6	4	15	23	32	36	40	62	64	75	85	89
1768	7	4	15	22	33	36	40	62	64	74	83	86
1768	8	4	15	22	33	37	41	62	66	76	86	89
1768	9	4	14	20	33	37	41	63	66	76	86	90
1768	10	4	14	21	34	37	41	62	66	76	86	88
1780	1	4	14	20	32	36	40	58	61	72	82	82
1780	2	5	14	20	33	37	42	59	64	73	85	84
1780	3	6	14	19	35	39	44	59	62	74	86	83
1780	4	4	14	20	32	34	39	57	60	71	84	81
1780	5	4	13	19	33	39	42	55	60	71	83	80
1780	6	6	14	19	33	37	42	57	62	70	82	80
1780	7	6	14	19	35	40	44	59	62	73	86	84
1780	8	5	14	20	33	38	42	59	61	74	84	83
1780	9	6	14	20	35	39	44	60	63	75	88	86
1780	10	5	14	19	34	39	42	59	62	74	86	83
1781	1	5	16	22	34	40	44	60	64	76	86	90
1781	2	5	16	22	35	40	45	65	69	79	88	92
1781	3	5	16	22	34	40	44	64	67	77	87	90
1781	4	5	16	23	34	40	44	65	69	79	87	93
1781	5	5	16	22	34	39	43	61	64	75	84	88
1781	6	5	16	22	34	39	44	61	64	75	83	90
1781	7	6	16	22	36	41	46	62	68	79	87	92
1781	8	5	16	22	32	37	42	62	65	76	85	90
1781	9	5	16	22	34	39	43	62	66	76	85	90
1781	10	5	16	22	34	40	44	62	65	76	86	91
1782	1	4	15	21	35	40	44	60	62	76	84	89
1782	2	4	14	20	34	40	43	58	60	76	85	87
1782	3	4	14	21	33	38	42	59	61	76	83	87
1782	4	4	13	20	34	39	42	60	62	76	84	88
1782	5	4	13	20	34	40	43	58	60	76	84	86
1782	6	4	15	21	34	40	44	62	64	79	87	90
1782	7	4	14	20	34	39	43	59	61	74	83	86
1782	8	4	14	20	33	38	41	59	61	76	83	88
1782	9	4	14	20	34	39	43	59	61	76	84	87
1782	10	5	14	21	34	40	43	58	60	76	84	87
1835	1	6	16	22	37	43	48	65	70	82	93	95
1835	2	5	16	23	36	42	47	66	70	81	92	96
1835	3	6	16	22	38	43	47	64	69	79	88	94
1835	4	5	16	22	35	41	45	65	69	80	90	95
1835	5	5	16	23	37	42	47	68	73	82	92	98
1835	6	5	16	22	36	42	46	65	69	80	90	96
1835	7	5	17	23	37	42	46	67	72	81	91	98
1835	8	6	16	22	37	43	47	65	69	80	90	96
1835	9	6	16	22	36	43	47	65	69	80	89	94

Raw mandible data (continued)

SN	N	X1	X2	X3	X4	X5	X6	X7	X8	X9	X10	X11
1835	10	6	16	22	38	44	48	65	69	80	90	96
1836	1	4	15	22	36	41	45	64	67	82	91	94
1836	2	5	14	21	37	42	46	64	68	82	91	92
1836	3	5	15	22	38	42	48	64	68	84	95	96
1836	4	4	14	20	36	42	45	64	66	82	90	92
1836	5	4	14	21	36	42	46	64	67	83	92	94
1836	6	5	14	21	38	45	48	64	68	83	94	96
1836	7	4	15	22	35	40	45	64	66	83	90	93
1836	8	5	15	21	36	42	46	63	66	80	90	92
1836	9	4	14	21	36	40	64	64	67	83	92	93
1836	10	4	14	21	35	40	45	64	66	80	89	93
1838	1	6	14	20	37	43	46	61	66	80	90	93
1838	2	6	14	20	38	42	48	65	70	80	91	92
1838	3	7	14	20	38	41	48	65	70	80	91	92
1838	4	6	14	22	38	40	47	65	70	81	92	93
1838	5	6	15	20	37	40	46	62	66	78	88	91
1838	6	7	16	22	38	42	46	65	69	82	93	95
1838	7	6	14	21	38	42	47	62	67	81	92	94
1838	8	6	15	20	36	40	46	61	67	74	87	88
1838	9	6	14	19	36	40	44	59	64	77	88	90
1838	10	6	14	20	37	40	46	62	65	80	90	90
1839	1	5	14	20	35	40	44	60	64	76	87	86
1839	2	6	15	20	36	40	44	59	63	75	87	86
1839	3	6	15	22	35	40	44	61	66	75	88	86
1839	4	6	15	21	36	41	46	62	67	80	92	98
1839	5	6	15	21	34	38	43	60	64	76	86	86
1839	6	5	14	20	34	38	43	60	65	76	88	84
1839	7	6	14	20	35	39	43	62	66	77	90	87
1839	8	6	15	22	36	41	46	61	67	78	90	86
1839	9	7	14	20	34	38	44	60	64	74	84	84
1839	10	6	15	22	36	42	46	63	67	79	90	88
1841	1	5	16	23	35	40	46	65	69	80	92	94
1841	2	6	16	21	36	43	47	64	69	78	89	92
1841	3	6	16	22	38	44	48	66	71	80	91	96
1841	4	6	16	22	37	43	47	64	69	78	90	94
1841	5	5	16	23	36	41	46	64	69	79	90	93
1841	6	7	16	22	39	44	48	64	69	79	91	94
1841	7	6	17	24	36	42	47	65	69	80	93	95
1841	8	6	16	22	37	43	48	65	70	80	92	96
1841	9	6	16	22	39	46	50	65	69	80	92	94
1841	10	6	16	22	37	42	48	66	71	80	92	96

References

Afifi, A. A. and S. P. Azen (1979). Statistical Analysis. A computer-oriented approach. Academic Press, New York.

Blackith, R. E. and R. A. Reyment (1971). Multivariate Morphometrics. Academic Press, New York.

Buhler, S. and R. Buhler (1979). PSTAT 78 User Manual. PSTAT Inc. P.O. Box 285. Princeton N.J. 08540 U.S.A.

Cooley, W. W. and P. R. Lohnes (1971). Multivariate Data Analysis. John Wiley and Sons, New York.

Dixon, W. J. and M. B. Brown (1979). BMDP Biomedical Computer Programs. P-series. University of California Press, Berkeley.

Festing, M. F. W. (1972). Mouse strain identification. *Nature*, **238**: 351.

Festing, M. F. W. (1973a). Mouse strain identification by mandible analysis. *In* A. Spiegel (ed.) The Laboratory Animal in Drug Testing. 5th ICLA Symposium. Gustav Fischer Verlag, Stuttgart, pp. 105–113.

Festing, M. F. W. (1973b). A multivariate analysis of subline divergence in the shape of the mandible in C57BL/Gr mice. *Genet. Res. Camb.*, **21**: 121.

Festing M. F. W. (1974a). Footpad bristles: a convenient metric character in mice. *Genet. Res. Camb.*, **24**: 315.

Festing, M. F. W. (1974b). Genetic reliability of commercially-bred laboratory mice. *Lab. Animals*, **8**: 265.

Festing, M. F. W. (1974c). Genetic monitoring of laboratory mouse colonies in the Medical Research Council Accreditation Scheme for the suppliers of laboratory animals. *Lab. Animals*, **8**: 291.

Festing, M. F. W. (1976). Effects of marginal malnutrition on the breeding performance of inbred and F1 hybrid mice—a diallel study. *In* Th. Antikatzides, S. Erichsen and A. Spiegel (eds.) The Laboratory Animal in the Study of Reproduction. Gustav Fischer Verlag. Stuttgart, pp. 99–114.

Festing, M. F. W. (1979). Inbred Strains in Biomedical Research. The Macmillan Press, London.

Festing, M. F. W. and D. P. Lovell (1979). Routine genetic monitoring of commercial and other mouse colonies in the UK using mandible shape; five years of experience. *In* A. Spiegel, S. Erichsen and H. A. Solleveld (eds.) Animal Quality and Models in Biomedical Research. Gustav Fischer Verlag, Stuttgart, pp. 341–348.

Gruneberg, H. (1952). The Genetics of the Mouse. 2nd ed. Martinus Nijhoff, The Hague.

Lovell, D. P., R. K. Archer, J. Riley and R. K. Morgan (1981). Variation in haemotological parameters among inbred strains of rat. *Lab. Anim.*, **15**: 243.

IV. CYTOGENETIC TECHNIQUES FOR GENETIC MONITORING

Chromosomes are good markers for the genetic monitoring of experimental animals. Identification of mouse and rat strains is now possible through karyotype analyses that use various band staining techniques. We can also observe chromosomes in living animals by means of culture techniques. In this section, techniques will be introduced for chromosome observations in the mouse and rat.

Preparation of Slides for Chromosome Observation

Bone marrow technique

Through the use of the bone marrow technique, slides for chromosome observation can be made directly from the animal without necessitating any procedures for tissue culture. The technique is performed as follows:

a) An adult mouse is injected with 0.1 ml of 0.1% (w/v) colchicine or colcemid (0.4 ml in the adult rat) in the peritoneal cavity, and killed one to two hours later.

b) The femur is cut off (Fig. 24a), and the bone marrow cells from the shaft of the bone are placed in a centrifuge tube in phosphate buffered saline (PBS) by means of a syringe (Fig. 24b).

c) The suspension is gently centrifuged at 1,200 rpm for 5 minutes, after which the PBS is discarded.

d) About 3 ml of 1% (w/v) sodium citrate is added to the tube through a pipette to make a suspension of bone marrow cells (Fig. 24c).

e) The suspension is left at room temperature for 10 minutes.

f) Five ml of Carnoy 3:1 (absolute alcohol : gracial acetic acid) is gently added to the tube, gently mixed with sodium citrate solution, and left for 10 minutes (Fig. 24d).

g) The suspension is centrifuged and the solution is removed. Three ml of Carnoy solution is added and the mixture is left for 10 minutes.

h) After another centrifugation, the solution is discarded and about 0.2 ml Carnoy solution is added.

i) A drop of the fixed material is placed on a wet slide kept in 50% alcohol, dried quickly over a spirit or gas flame (Fig.

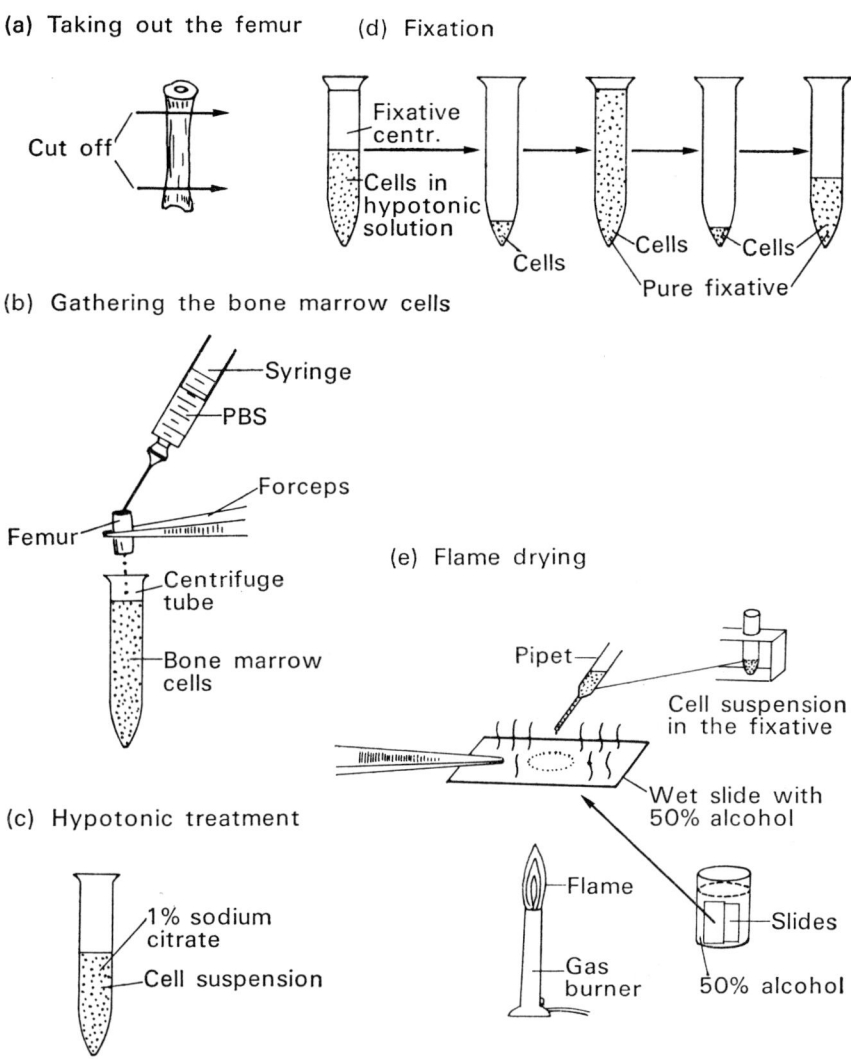

(a) Taking out the femur

Cut off

(d) Fixation

Fixative centr.

Cells in hypotonic solution

Cells

Cells

Cells

Pure fixative

(b) Gathering the bone marrow cells

Syringe

PBS

Forceps

Femur

Centrifuge tube

Bone marrow cells

(e) Flame drying

Pipet

Cell suspension in the fixative

Wet slide with 50% alcohol

(c) Hypotonic treatment

1% sodium citrate

Cell suspension

Flame

Gas burner

Slides

50% alcohol

FIG. 24. Procedure for the flame drying technique used for bone marrow cells for chromosome observation of rodents.

24e), and then stained with acetic orcein or 2% Giemsa solution (Fig. 25). For band staining, the slide is dried without the use of a flame.

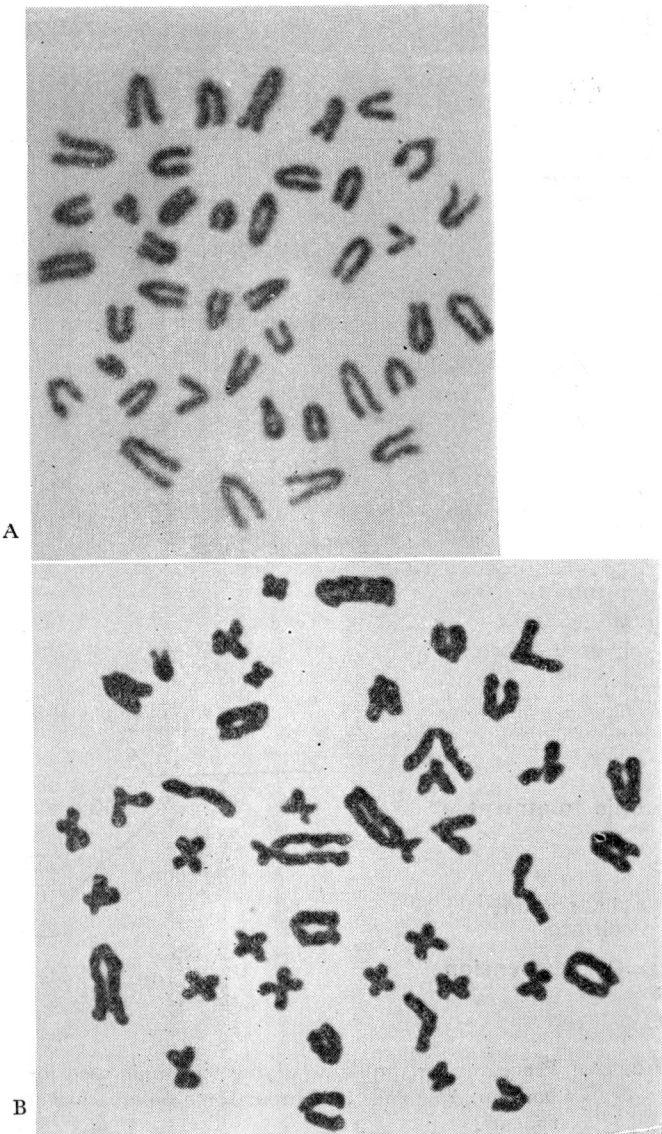

FIG. 25.   Bone marrow chromosomes of the mouse and rat.
A: mouse metaphase by acetic orcein staining;
B: rat metaphase by conventional Giemsa stain-
ing.

Tail culture technique

The tail culture technique for the observation of chromosomes in cultured cells of the tail can be used in the mouse or any other rodent with a long tail. The technique does not require the sacrifice of the animal (Yosida, 1980).

a) The end 5 cm of the tail are cleaned with alcohol, and a piece is cut off approximately 2–3 cm from the tip.

b) In a sterilized room, the skin of the tail is stripped with a forceps. The stripped tissue is then washed in saline (PBS).

c) The tissue is minced into small pieces with a scissors and placed in a petri dish. The small pieces of tissue are placed on a culture dish after the extra moisture has been absorbed by sterilized blotting paper.

d) The tissue pieces are left on the culture dish for about 10 minutes to fix.

e) Five ml of culture medium (Eagle MEM+20% fetal calf serm) is added, and the culture is incubated in a CO_2-incubator at 38°C.

f) About 5 days after cultivation, during which time the cells should have multiplied sufficiently to yield a harvest, the culture medium is refreshed.

g) About 24 hours later, two drops of colchicine or colcemid (final concentration 0.5–1.0 $\mu g/ml$) are added to the culture dish and it is left for one hour.

h) The culture medium is poured into the centrifuge tube, and the culture dish is washed with PBS.

i) Two ml of trypsin solution (0.025%) is added to the dish and it is left for 5 minutes in the 38°C incubator.

j) A suspension of the harvested cells is then poured into the centrifuge tube.

k) The cell suspension is centrifuged and treated with hypotonic 1% sodium citrate.

Subsequently the routine cytogenetic technique used for bone marrow cells is applied in preparing the slides. If the initial attempt at cultivation of the cells fails, a second piece of tail from the same rat can be obtained and treated in the same way.

Ear culture technique

If the tail culture technique fails or cannot be used, a piece of ear tissue is used for cultivation.

a) The ear is cleaned with alcohol and a piece is cut off.

b) The ear piece is washed with PBS (including kanamycin) in a sterilized Petri dish.

c) The ear piece is minced into small pieces. Further procedures are the same as those for tail cultivation.

Staining Techniques

Conventional staining with Giemsa solution

Slides made from bone marrow cells or cultured cells are stained directly for 2–3 minutes with 2% Giemsa solution. The slide is rinsed with tap water, dried, and examined microscopically.

Differential staining techniques

A. G-band staining technique

i. Trypsin treatment

The trypsin digestion procedure is commonly used for G-band in bone marrow and cultured cells of rodents. The following is a modification of the Seabright (1971) method (Yosida, 1980):

a) A slide is made from cultured cells of tail or ear, or bone marrow cells by air drying, and left overnight.

b) The slide is treated with 0.025% trypsin solution for 30 to 40 seconds,

c) washed for 5 seconds in phosphate-buffered saline (PBS) at pH 6.8,

d) stained for 2–3 minutes with Giemsa solution diluted with Sörensen's buffer (pH 6.8), and

e) rinsed with tap water and dried (Figs. 26 and 27).

ii. SDS treatment (Yosida and Sagai, 1972)

a) Chromosome preparation is made by air drying.

b) The slide is immersed in 0.1% (w/v) SDS (sodium dodecyl sulfate or sodium lauryl sulfate) in $2 \times SSC$ for a few seconds at room temperature. It is then washed with tap water and stained for about 10 minutes in Giemsa solution diluted in Sörensen's buffer (pH 6.8) (Yosida and Sagai, 1972)

155

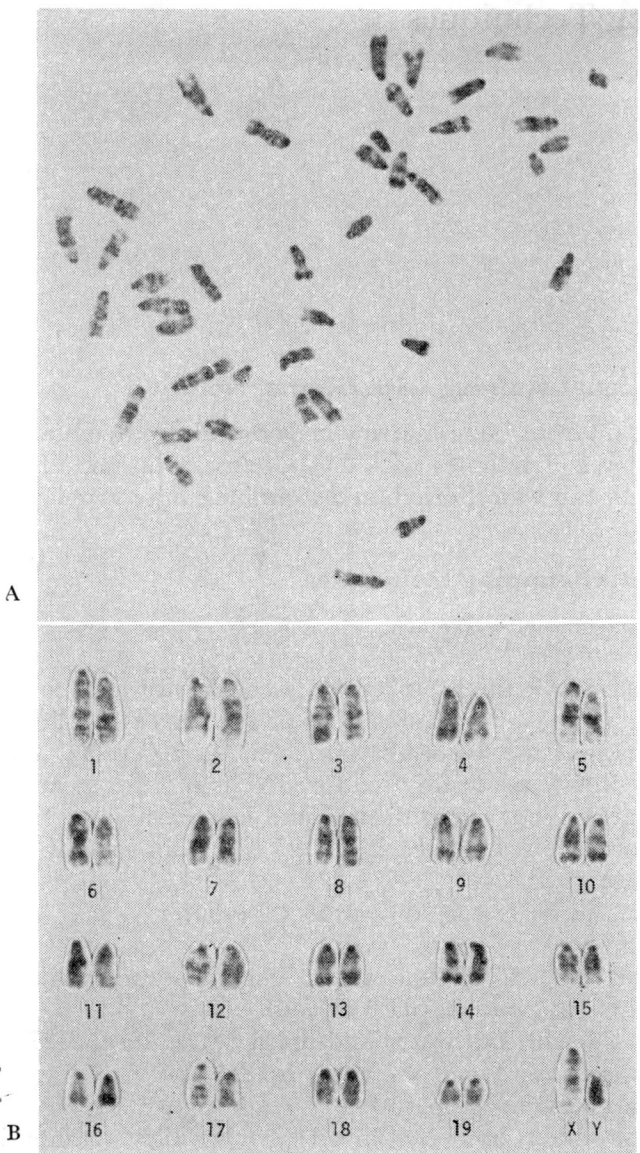

FIG. 26.   G-band staining by trypsin treatment of the mouse.
A: metaphase; B: karyotype. The chromosome
arrangement follows the Standard Karyotype of
the Committee on Standard Nomenclature for
Mice (1972).

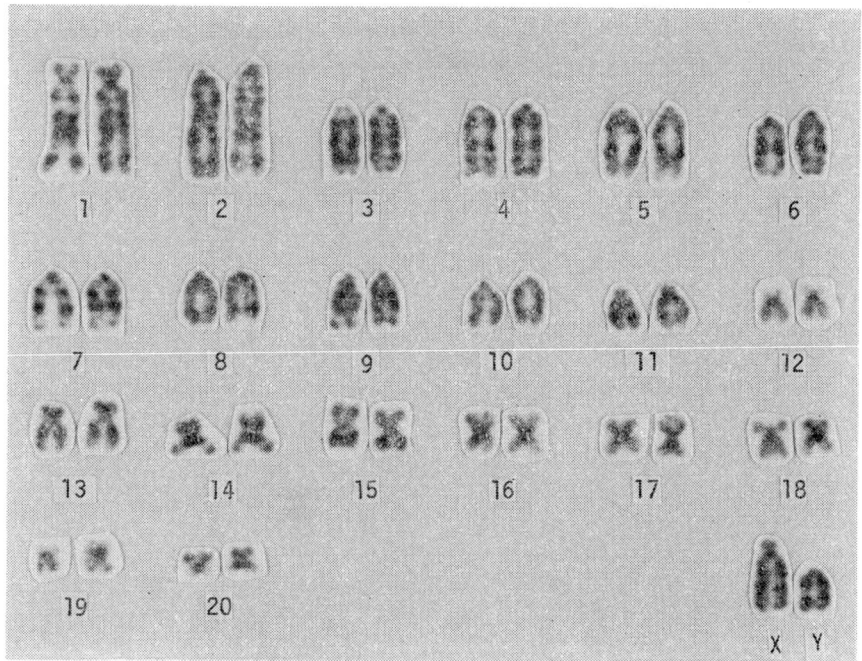

FIG. 27. G-band karyotype of the rat (*R. norvegicus*). Chromosomes are arranged according to the system of the Committee for a Standardized Karyotype of *Rattus norvegicus* (1973).

B. C-band staining technique

Sumner's technique (1972) is used, although the procedure is slightly modified (Yosida and Sagai, 1975).

a) The slides are made by the usual procedure from cultured cells or bone marrow cells by air drying.

b) The slides are immersed for 30 minutes in 0.2 N HCl,

c) treated with 0.5% Ba(OH)₂ for 5–6 minutes at 60°C,

d) immersed in 2 × SSC for 30 minutes, and

e) washed in tap water and stained with 3% Giemsa solution with phosphate buffer (pH 6.8) (Fig. 28).

C. Serial G- and C-band staining technique

Identification of chromosome pairs is difficult from C-band staining. Therefore, to identify chromosome pairs from C-banding,

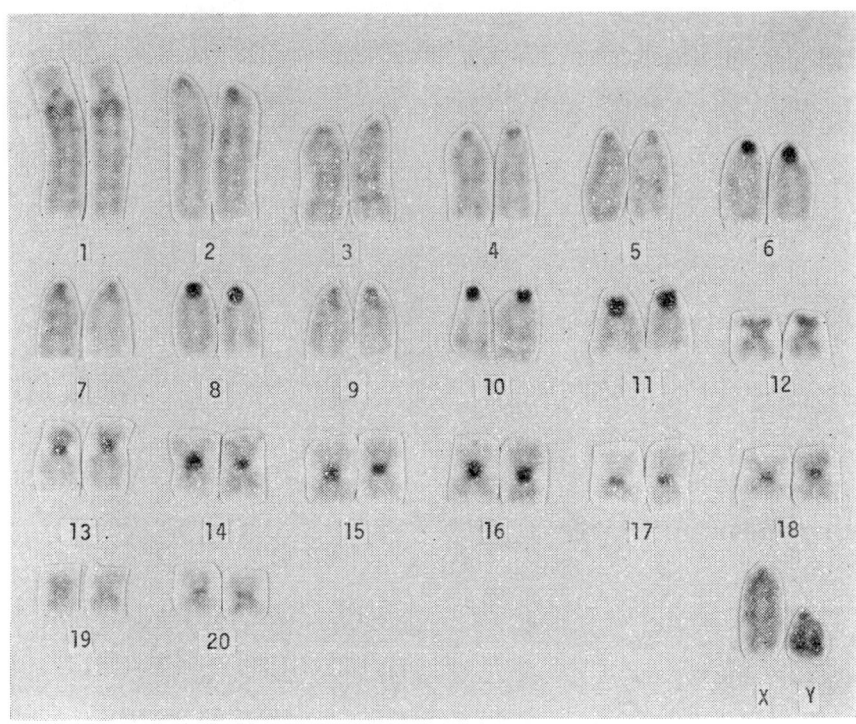

FIG. 28. C-band karyotype of the rat. C bandings differ
depending on the strain of the rat.

serial G- and C-band staining is applied (Yosida and Sagai, 1975).
  a) The slide is stained by the G-band technique described
      above. From this slide, photomicrographs of metaphase cells
      showing good G-bands are taken.
  b) The slide is then immersed in acetic acid (50%) for a few
      seconds to remove the G-band stain.
  c) After the slide is washed in tap water, C-band staining as
      described above is applied.
  d) Photomicrographs are again taken of C-band metaphases
      and compared with the G-band photographs taken previ-
      ously.

D. Q-band staining technique

Quinacrine fluorescence banding (Q-band) techniques have
been modified by several authors, but the standard procedures

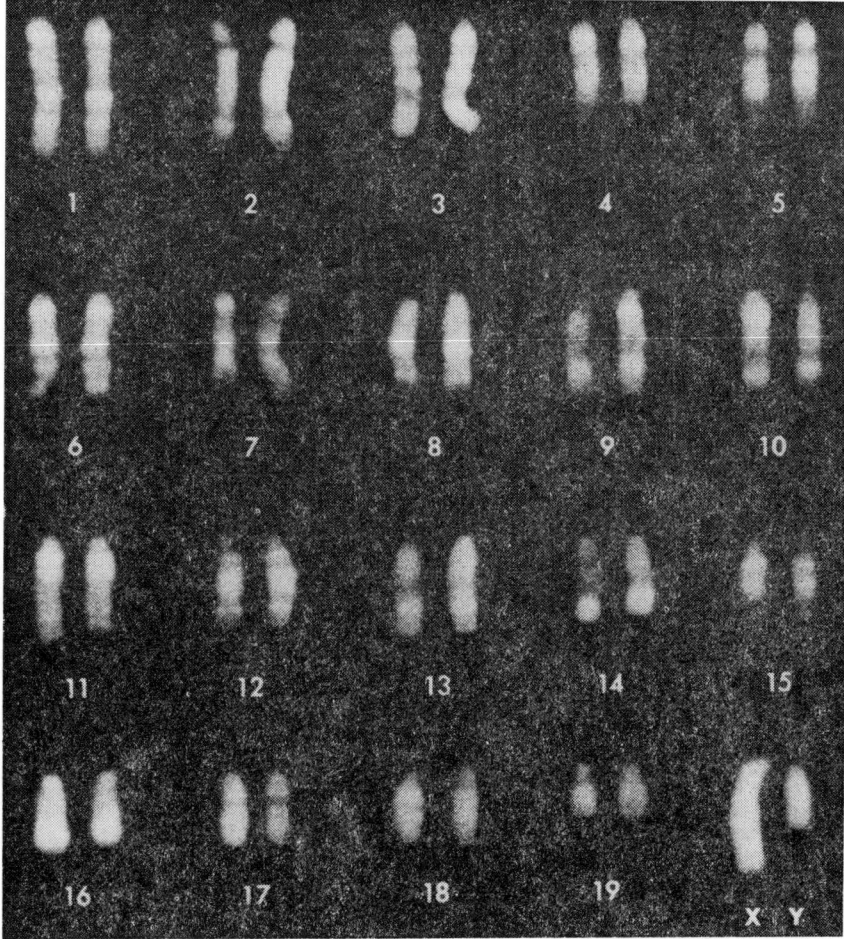

FIG. 29.   Q-band karyotype of the mouse (from the Committee on Standard Nomenclature, 1972).

described by Caspersson *et al.* (1971) will be introduced here. The
banding patterns obtained with this method correspond to those
obtained by G-banding.

   a)   The slides prepared by air drying are transferred from
      alcohol steps and buffer (MacIlvaine's disodium phosphate-
      citrate acid buffer, pH 7.0) into staining solution.

   b)   An aliquot of quinacrine mustard dihydrochloride in aqueous solution is added to the buffer to give a final concentration of 50 μg/ml of the fluorochrome in the staining solution. After staining for 20 minutes at 20°C, the slides are washed three times in buffer and sealed with a cover slip in buffer.

c) The slides thus prepared are examined under the fluorescence microscope, and the chromosomes showing fluorescent bands are photographed (Fig. 29).

E. *Differential staining of nucleolus organizers (NORs)*

To stain the NORs of chromosomes differentially, a modification of the Ag-As staining technique of Goodpasture and Bloom (1975) was used as follows (Yosida, 1980):

i. Three solutions are prepared:

a) Ag solution:

$$AgNO_3 \qquad 10 \text{ g}$$
$$\text{Distilled water} \quad 20 \text{ ml}$$

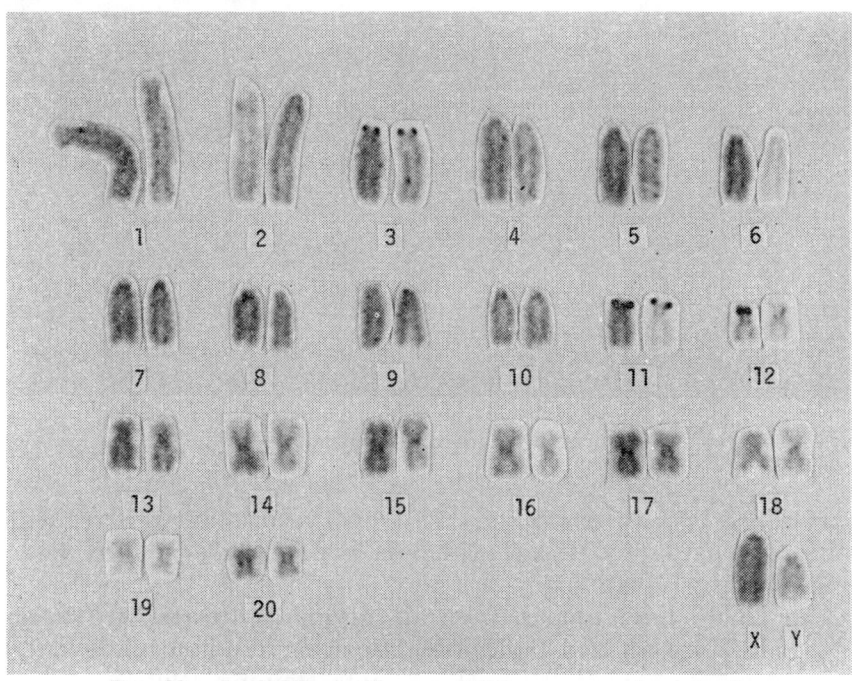

Fig. 30. Ag-NORs in the rat. NORs are seen in pairs 3, 11, and 12.

b) As solution:
          $AgNO_3$ 8 g
          Distilled water   10 ml
          NH_4OH 15 ml

c) 3% formalin:
          Formalin   3 ml
          Distilled water   97 ml

The solution is adjusted to pH 3–4 with formic acid.

ii. The staining procedure is as follows:

a) Five drops of Ag solution are pipetted onto a slide, prepared by air drying, and covered with cover glass.

b) The slide is placed about 20 cm below a photo flood (WEST 500 W bulb) for 10 to 15 minutes.

c) The slide is [rinsed [with tap water, separated from its cover glass, and dried.

d) Four or 5 drops of As solution are pipetted onto the slide, followed by 4 or 5 drops of 3% formalin, and the cover glass replaced.

e) Staining progress is checked under the phase-contrast micro-scope. When the surface of the cells reaches a yellow color, the slide is rinsed in tap water and the cover glass is taken off.

f) The slide is stained with 4% Giemsa solution with phosphate buffer for 3 to 4 minutes, washed in tapwater, and dried (Fig. 30).

Karyotype Analysis

In karyotype analysis photographs of mitotic figures are taken and enlarged to an original magnification of 3,000 to 4,000 times (Fig. 25). The chromosomes are cut out from the prints and arranged in order of decreasing size into groups of chromosome types. In conventional staining, the arrangement of chromosomes with homologous pairs is difficult in the mouse and rat because they have many similar acrocentric or metacentric chromosome pairs. Through differential staining (G or Q), the identification of all chromosome pairs is possible (Figs. 26, 27 and 29) in these animals. G-band karyotypes of the mouse and rat are shown in Figs. 26 and 27. They are arranged according to the system used by the Committee on Standard Nomenclature for Mice (1972) and the Committee for a Standardized Karyotype of *Rattus norvegicus* (1973). In karyotyping from C bands and NORs, double stainings such as G to C bands or G to NOR stainings are desirable to determine homologous chromosome pairs and chromosome pair numbers.

References

Caspersson, T., G. Lomakka, and L. Zech (1971). The 24 fluorescence patterns of the human metaphase chromosomes: Distinguishing characters and variability. *Hereditas,* **67**: 89–102.

Committee on Standard Genetic Nomenclature for Mice (1972). Standard Karyotype of the Mouse, *Mus musculus. Jour. Hered.,* **63**: 69–72.

Committee for a Standardized Karyotype of *Rattus norvegicus* (1973). Standard Karyotype of the Norway rat, *Rattus norvegicus. Cytogenet. Cell Genet.,* **12**: 199–205.

Goodpasture, C. and S. E. Bloom (1975). Visualization of nucleolar organizer regions in mammalian chromosomes using silver staining. *Chromosoma* (Berl.) **53**: 37–50.

Seabright, M. (1971). A rapid banding technique for human chromosomes. *Lancet ii*: 971–972.

Sumner, A. T. (1972). A simple technique for demonstrating centromeric heterochromatin. *Exp. Cell Res.,* **75**: 304–306.

Yosida, T. H. (1980). Cytogenetics of the Black Rat: Karyotype Evolution and Species Differentiation. Univ. Tokyo Press, Tokyo, and Univ. Park Press, Baltimore.

Yosida, T. H. and T. Sagai (1972). Banding pattern analysis of polymorphic karyotypes in the black rat by a new differential staining technique. *Chromosoma* (Berl.), **37**: 387–394.

Yosida, T. H. and T. Sagai (1975). Variation of C-bands in chromosomes of several subspecies of *Rattus rattus. Chromosoma* (Berl.), **50**: 283–300.

General References

Ashwood-Smith, M. J. and J. Farrant (Eds.) (1980). Low Temperature Preservation in Medicine and Biology. University Park Press, Baltimore.

Carpenter, P. L. (1975). Immunology and Serology (3rd ed.). W. B. Saunders Company, Philadelphia.

Elliot, K. and J. Whelan (Eds.) (1979). The Freezing of Mammalian Embryos. Ciba Foundation Symposium 52 (New series). Excepta Medica, Amsterdam.

Festing, M. F. W. (1979). Inbred Strains in Biomedical Research. Macmillan Press. Ltd. London.

Foster, H. L., J. D. Small, and J. G. Fox (Eds.) (1981). The Mouse in Biomedical Research. Vol. I. History; Genetics and Wild Mice. Academic Press, New York.

Foster, H. L., J. D. Small, and J. G. Fox (Eds.) (1983). The Mouse in Biomedical Research. Vol. III. Normative Biology; Immunology and Husbandry. Academic Press, New York.

Green, M. C. (Eds.) (1981). Genetic Variants and Strains of the Laboratory Mouse. Gustav Fischer Verlag, Stuttgart.

Gotze, D. (Ed.) (1977). The Major Histocompatibility System in Man and Animals. Springer Verlag, Berlin.

Klein, J. (1975). Biology of the Mouse Histocompatibility-2 Complex. Springer Verlag, Berlin and New York.

Mühlbook, O. (Ed.) (1976). Basic Aspects of Freeze Preservation of Mouse Strains. Gustav Fischer Verlag, Stuttgart.

Zeilmaker, G. H. (Ed.) (1981). Proceedings of the Workshop on Embryo Storage and Banking in Laboratory Animals. Gustav Fischer Verlag, Stuttgart.

SUPPLEMENT

Summary of the ICLAS/ICREW Workshop on Preparation of an International Manual on Genetic Monitoring for Inbred Strains of Mice Used in Cancer Research.

Introduction

On behalf of the Organizing Committee and ICLAS, I would like to briefly give the background to this Workshop. This supplement also contains a Special Lecture stressing the need for an international standard for genetic monitoring, reports on the present status of genetic monitoring in various countries, and the conclusions approved by all the Workshop participants.

This Workshop was realized at the request of ICLAS with the aid of the ICREW Program—the International Cancer Research Workshop Program administered by the International Union Against Cancer (UICC). As you may know, ICLAS was founded in 1956 as ICLA—International Committee on Laboratory Animals. In 1979 the new name ICLAS—International Council of Laboratory Animal Science—was adopted.

One of the aims of ICLAS is to promote standardization in laboratory animal science, and since 1970, several reference centers have been established to achieve this aim. Examples include centers for histocompatibility testing and reference centers for biochemical markers for mice and rats from the genetic standpoint; virus reference centers for rodents from the microbiological standpoint; and reference centers for nude mice. These centers have been entrusted to various research institutions or researchers. ICLAS has established basic policies, but new policies still have to be formulated in accordance with]ICLAS aims. The next step is a genetic and microbiological monitoring center system.

The Governing Board of ICLA asked the National Members from the United States and Japan, Dr. Hopla and myself, to carry out a basic study of such a monitoring center system. In the summer of 1978, Dr. Hoffman came to Japan and carried out basic investigations on monitoring items and methods; and through close consultation, a basic draft was drawn up. Based on this draft, we started actual genetic monitoring in Japan. At the 1979 ICLAS General Assembly I gave a report on the results of this monitoring, and Dr. Hoffman presented a paper at the symposium.

The ICLAS Governing Board adopted a new Monitoring Center System Program for Laboratory Animals and appointed me as chairman of the Program.

Dr. Odartchenko recognized the importance of such monitoring and suggested that it would be useful for ICLAS to hold a workshop on genetic monitoring with the assistance of the ICREW Program of UICC. This Workshop is the result.

This Workshop has two important tasks. The first is to prepare an international manual on genetic monitoring of inbred strains of mice.

The second task concerns ICLAS policy. Although ICLAS has established several reference centers to promote standardization in laboratory animal science, their number is still not sufficient. Therefore, this Workshop is the first definitive step in the promotion of standardization by ICLAS.

Once genetic monitoring centers are established in each part of the world as the outcome of this Workshop, the final goal of international quality standardization for laboratory animals can be achieved.

Tatsuji Nomura

Chairman of the Workshop
of an International Manual
on Genetical Monitoring for
Inbred Strains of Mice Used
in Cancer Research

Special Lecture: The Need for an International Genetic Quality Monitoring System

The search for high quality standards in laboratory mice used for cancer research has a long history.

The Swiss mouse, to take a familiar example, now invading research laboratories the world over, can be clearly traced back to 2 males and 5 females given to C. Lynch and smuggled by her into the Rockefeller Institute in New York. This gift, in 1926, was from an early pioneer of our present Cancer Institute, A. de Coulon, then chief of the experimental division, who had a colony of several thousand mice which he used for his work on tar- and arsenic-induced cancer. The original Lausanne colony has now been extinct for a long time, but the circle is completed as our present Swiss mice are being sent back to us from descendants of those early travellers.

Now, in 1980, the total number of inbred strains of mice and of properly identifiable substrains exceeds 300. Their use in cancer research extends from immunology and virology to basic questions of environmental carcinogenesis, to the study of genetic factors in susceptibility to cancer, and to the analysis of the role of nutritional and hormonal factors, to mention only some fundamental fields. No simple answers are expected, particularly in view of the ever-growing importance of basic problems such as the level at which cellular control mechanisms operate in Metazoans, be it at a translational level or at the level of enzyme synthesis, as well as the permanent increase in knowledge on genetic polymorphisms in mice.

The introduction of inbred strains for cancer research has been properly given the same importance as that of the analytical balance in chemistry. But an analytical balance needs periodical service and standards of comparison of the accuracy of measurements. A system of efficient genetic monitoring would render this

169

service by allowing detection of differences between inbred sub-lines, should they be due to contamination from out-crossing, incomplete inbreeding, or mutation, as formulated by D. W. Bailey.

One of the greatest and most useful benefits to be gained from a monitoring system, allowing control of the genetic integrity of inbred strains, is comparison of results in space and time. By this means, it will be possible to discover and to correct anomalies, thus helping scientists and granting agencies in the orientation of research tasks and in the distribution of resources as well as encouraging competition between research groups.

One of the objectives of the International Cancer Research Workshop (ICREW) program, administered by the International Union against Cancer (UICC), is "to evaluate or to plan methods for overcoming some particular obstacles or for resolving a specific disagreement impeding further progress." There is no doubt that the task the organizers and the participants of this Workshop have set forth corresponds to this objective, and for this reason it has been granted ICREW sponsorship. I see, however, an additional significance resulting from this recognition: the truly international character of the need which, I hope, your work will fulfill.

With greetings from UICC, I wish you a most successful workshop and look forward to the production of a genetic monitoring manual internationally indispensable for cancer researchers in their work with mice.

N. Odartchenko
Swiss Institute for Experimental
Cancer Research
Epalinges/Lausanne, Switzerland

Current Status of Genetic Monitoring

In the Federal Republic of Germany

In the Federal Republic of Germany approximately 1,000,000 inbred mice are used in biomedical research. Fifty percent are bred at research laboratories or at central laboratory animal facilities of the universities. All other mice are bred and distributed by four national and four foreign laboratory animal breeders. A functioning genetic monitoring scheme is restricted to a very few laboratories.

The system employed at the Central Institute of Laboratory Animals is based on two principles. The first principle is high standard colony management. This includes a distinct structure of the breeding colony. The stem line is maintained in a modified parallel line system at a limited size with a slowed-down generation turnover rate and with high standard animal maintenance.

The second principle is genetic control of the breeding colonies applied at set intervals. This control includes physical examination of the animals, evaluation of parental reproductive data, and genetic profiling for as many mono- and polygenetic markers as possible to define the strain-specific standards. The selection of markers should guarantee that they are equally distributed throughout the genome, but special properties of certain strains/ mutants should also be considered. Four different control measures are applied: (1) Standard observation procedures are undertaken: general examination as well as pigmentation and pelage variation. (2) Immunogenetic markers such as lymphocyte antigens and histocompatibility antigens are assayed by serotyping and skin grafting. (3) Several biochemical markers are checked by starch-gel electrophoresis. (4) A mandible analysis which tests osteometric traits is also used. So far 25 monogenetic markers plus polygenetic systems are included in the monitoring scheme. The list of markers is considered neither complete nor final. However, the inclusion of other markers depends on the ease with which a new method or technique can be established. It depends on funding and technical and personnel equipment of the monitoring laboratory.

Freeze preservation of embryos has been begun. It is considered to be the optimal technique to maintain inbred strains genetically constant.

Hans J. Hedrich
Zentralinstitut für Versuchstiere
F.R. Germany

In France

There are three main sources of supply for inbred mice in France: one government institution and two private firms.

The government institution maintains and produces about 23 inbred strains of mice, most of them recently introduced from the Jackson Laboratory. Moreover, 27 recombinant and congenic resistant inbred strains from various origins are also maintained.

One of the private firms produces 4 inbred strains of mice related to those of the National Institute of Health; the other firm, 6 inbred strains of various origins.

The government institution has established a routine genetic quality control program of inbred strains, using the coat-color, skin grafting and biochemical marker methods. Since 1976 it has been designated as an ICLAS Reference Center for Biochemical Markers and therefore provides biochemical testing services for other laboratories as well as training and information.

Although the private firms actually monitor the genetic control of their own inbred strains independently of the government institution, direct and indirect evidence suggests that genetic divergence of sublines exists to some extent for commercial inbred strains currently available in France.

R. Moutier
Centre de Sélection et
d'Elevage d'Animaux
de Laboratoire, C.N.R.S.
France

In the United Kingdom

Although some skin grafting of inbred strains of mice has been carried out in the U.K. for many years, most colonies are maintained entirely without tests of authenticity. As the use of inbred strains has increased, genetic contamination has become of increasing importance, leading in some cases to serious loss of research effort. The development of the mandible shape technique of strain identification[1] made it possible to offer routine genetic monitoring as part of the Accreditation Scheme for commercial breeders in the U.K., administered by the MRC Laboratory Animals Centre.

Rules for this genetic monitoring scheme were published in 1976.[2] Any breeder who wished to enter had to provide details of his colonies, and he had to satisfy the Laboratory Animal Centre that he was maintaining his colonies correctly. He also had to pay about £30 per strain per year to cover the direct costs of the scheme, and provide samples of animals for testing when so requested.

A total of 38 strains and stocks have now been monitored for an average of 3 1/2 years per strain.

A total of 341 samples of 10 mice have been processed to date. Of these, 19 (6%) were judged to be "doubtful." This led to the identification of at least five cases of genetic contamination which were verified using other techniques. Most of the problems involved a single commercial breeder, who has now taken steps to improve the genetic quality of his stock.

The mandible shape technique has proved useful for such routine monitoring. It is quick, economical and reasonably reliable. However, the technique is not entirely satisfactory when genetic contamination is suspected. Other techniques are needed in such cases. These techniques can include skin grafting, the study of biochemical markers, test matings to reveal hidden coat color loci in the case of albino strains, and immunological methods including the use of polyvalent alloantisera.[3] Choice of method depends on individual circumstances.

The future of the Genetic Monitoring Scheme is now under review. A centralized governmental scheme is expensive and lacks flexibility. Breeders should therefore be encouraged to develop their own in-house methods. These might be backed up by a central laboratory, which could act as arbiter in cases of dispute and could offer a wider range of techniques than would be used for routine work. Further research on simplified methods of monitoring is also needed. For example, calculations for the mandible shape method have now been programmed into a handheld T159 calculator, which costs only about £200. Such equipment should be easily available to the average animal house. If such simplified methods can be further developed, it may be possible for the animal technician to carry out his own routine monitoring within the animal house, provided he is suitably backed up when difficulties arise.

1. Festing M.F.W. (1972) *Nature* **238**: 351.
2. The Genetic Monitoring (GM) Scheme. MRC Laboratory Animals Centre, Carshalton, Surrey, U.K. (1976).
3. Festing M.F.W. and Totman, P. (1980) *Laboratory Animals* **14**: 173.

Michael F. W. Festing
MRC Laboratory Animals Centre
U.K.

In Japan

Inbred strains of mice in Japan are maintained by many organizations. For example, the C3H strain was maintained by 61 organizations in 1974.

In 1978, we planned to establish a genetical monitoring system and to test the genetic quality of inbred mice, with the support of the Ministry of Education, Science and Culture.

After selecting 20 gene loci for the monitoring and establishing checking methods, we tested a total of 111 substrains of 10 inbred strains maintained in 24 organizations including 16 universities, 5 research institutes and 3 animal breeders.

Genetic polymorphisms were observed in eight out of 111 substrains. In one substrain each of C57BL/6, KK, DBA/2 and NZB, three to eight loci showed polymorphisms. This result led to the conclusion that genetic contamination had occurred in these substrains. Also, genetic variations among substrains were observed in C3H, AKR, CBA and NZB. Some of these substrains might be caused by inbreeding after genetic contamination or by taking one for another. We propose that these doubtful substrains be renewed to the defined strains or that the strain name be changed.

On the basis of the above work, a genetic monitoring system has almost been completed in Japan, and Dr. Nomura organized a monitoring center in the Central Institute for Experimental Animals which includes both genetical and microbiological checking.

Kozaburo Esaki
Central Institute for Experimental Animals
Japan

In Poland

The main source of mice for immunological and oncological research in Poland is two Institutes which coordinate all projects in the above-mentioned fields of studies. The animals in both Institutes represent the controlled conventional status of health and are registered with the Committee on Standardized Genetic Nomenclature for Mice as follows:

1. "Iiw" The Ludwik Hirszfeld Institute of Immunology and Experimental Therapy, Wrocław, Poland
2. "W" Maria Skłodoska-Curie Memorial Oncological Institute, Warsaw, Poland

In both Institutes the breeding nuclei of 27 inbred strains, 3 outbred and 18 congenic lines, as well as their production colonies are maintained. The genetically determined mice are given to other breeders (24 other research units) for production purposes limited to 2 generations. A routine genetic control program has been established to ensure that none of the inbred or congenic strains become genetically contaminated by nonstrain mating. At least once a year the following tests are performed: skin grafting among members of an inbred strain of F_1 hybrid, coat color test for allelism, identification of selected biochemical markers belonging to various linkage groups, and identification of immunological markers, mainly alloantigens of lymphocytes (Thy, Tl, Lyt, some specificities of H-2); natural incidence of tumors is routinely registered.

An activity of the Reference Center located at the Oncological Institute is the production of antisera for determination of H-s antigenic specificities controlled by the translocation chromosome T-190. Two congenic resistant lines (BALB. T190 and BALB. tf) were produced. In addition appropriate antisera were produced to determine specificities of $H\text{-}2^{bp}$ haplotype and to characterize 5 new congenic resistant lines (C3H.BN, BN.C3H, D2.BN, BN.D2, and B10.BN).

Czesław Radzikowski
Ludwik Hirszfeld Institute
of Immunology and Experimental Therapy
Poland

In the U.S.A.

At the Jackson Laboratory

The genetic quality of inbred strains maintained by the Animal Resources program of the Jackson Laboratory is monitored with two aims: (1) to assure that mice of a given inbred strain are free from genetic contamination, and (2) to eliminate observable mutations from inbred strains.

Continual monitoring of foundation stock mice selected as

breeders to propagate the strain is carried out with assays for biochemical markers. Assays for more than 20 different isoenzymes and other proteins are used, but generally from 4 to 8 such markers in addition to coat color are sufficient to distinguish a strain from all others maintained by the Jackson Laboratory.

Periodic monitoring is carried out to assess intra-strain histo-compatibility of all strains and to ascertain coat color alleles at the brown and agouti loci for albino strains. Histocompatibility is assayed by the exchange of tailskin grafts between randomly selected mice of the same sex within a strain. Coat color alleles are determined by making test crosses of albino mice with appropriate mice carrying the wild type allele at the albino locus.

Visible mutations are detected by animal husbandry personnel who are trained to watch for phenodeviants among the breeding colony of Animal Resources strains. Putative mutations are investigated and family lines are removed from the breeding colony if the observed deviation appears to be heritable.

Larry E. Mobraaten
The Jackson Laboratory
U.S.A.

At the National Institutes of Health
A variety of biochemical and immunological methods are currently being used to routinely monitor 45 inbred strains of mice and 24 inbred strains of rats maintained at the National Institutes of Health for their genetic integrity. The monitoring is based upon the concept of strain genetic profiles which are specific sets of genetic markers located on specific chromosomes. Therefore, genetic monitoring is the routine examination of specific sets of chromosomes (critical subset) which will uniquely discriminate one inbred strain from other inbred strains and substrains.

If the chromosomal analysis is considered in a probabilistic way, the probability that a sample from an inbred strain is not genetically contaminated can be calculated. When the inbred strains are first grouped by their phenotypic coat color, the critical subset of genetic markers can be determined to either differentiate an inbred strain from all other strains with the same coat color or differentiate it from a specific set of inbred strains. The computer is used to maintain all the records and genetic data associated with the routine monitoring, and also is used in the determination of the critical subset of genetic markers for strain certification.

Harold A. Hoffman
National Institutes of Health
U.S.A.

Conclusions

Inbred strains of laboratory mice produced as a result of 20 or more generations of brother × sister mating are now widely used in many areas of biomedical research, especially in immunology and cancer research. Inbred strains have many useful features which depend on the fact that all individuals of a single strain should be genetically identical at more than 99% of their genetic loci. Many inbred strains of mice are widely distributed, and have come to be regarded by some research workers as being an absolute standard on which research can be conducted using identical animals throughout the world. Thus, there are now over 300 colonies of "C57BL" mice maintained in colonies all over the world.

Recently, evidence has been accumulating which shows that many colonies of inbred strains of mice and rats are not genetically authentic. This is usually a result of accidental genetic contamination resulting from a mix-up in the animal house. A recent survey of 111 colonies of inbred mice maintained in Japan (T. Nomura, H. Hoffman, K. Esaki and T. Tomita) has shown that 8 colonies had evidence of recent genetic contamination, resulting in genetic segregation at one or more of the 20 genetic loci studied. Several other colonies showed subline deviation which could have been the result of genetic contamination in the past. In the U.K. a survey of 38 strains of mice over a five-year period revealed 5 cases of genetic contamination (1 contamination per 27 colony/years), and there were several cases where research projects had to be abandoned at considerable financial loss as a result of these contaminations (M.F.W. Festing). No comparable studies have yet been carried out in other countries, but there are several published papers, and many unconfirmed rumors that a similar situation can be found in the U.S.A. and elsewhere in Europe.

Methods of genetic monitoring aimed at identifying genetic contamination on a routine basis were discussed at a 1980 workshop in Tokyo, organized by Dr. T. Nomura of the Central Institute for Experimental Animals in Japan and sponsored jointly by the International Council of Laboratory Animal Science (ICLAS) and the International Cancer Research Workshop Program (ICREW). The participants were K. Esaki (Japan), M.F.W. Festing (U.K.), H.J. Hedrich (F.R. Germany), H.A. Hoffman (U.S.A.), L. Mobraaten (U.S.A.), R. Moutier (France), T. Nomura (ICLAS Japan), Cz. Radzikowski (ICLAS Poland), Y. Sakurai (UICC Japan), and T. Tomita (Japan). The main aim of the workshop was to prepare a manual outlining the methods that scientists and animal breeders may use if they wish to confirm that the animals they use are genetically authentic.

Unfortunately, it is not yet economically feasible (nor is it technically possible in many cases) to test each animal before it is used in an experiment. Attention must therefore be devoted to sampling the breeding colonies at various levels from the pedigree foundation stock to the final production animals. The technical methods range from immunological and biochemical methods based on the identification of alleles at a single locus known to be polymorphic in mice, to methods based on the simultaneous comparison of many loci such as skin grafting or the study of morphological characters such as the shape of the skeleton. The former methods are exquisitely precise but may be time-consuming and in some cases are technically complicated. The latter methods are often technically simple but may lack precision. Most monitoring schemes will authenticate the strain initially using a "genetic profile" of many biochemical and immunological markers, with routine monitoring being based on a restricted set of such loci or the less precise polygenic characters. The full genetic profile would then only need to be used at infrequent intervals or when trouble is encountered.

International cooperation on the standardization of laboratory animals is essential if experimental results are to be replicable in all countries. Genetic authentication of inbred strains of mice to ensure freedom from contamination is a small, but nevertheless important, step towards this goal. In the future it should be possible to achieve an even greater degree of genetic standardization by the development of banks of frozen mouse embryos, in which genetic changes as a result of contamination and genetic drift due to mutation can be almost entirely eliminated.

Michael Festing

Index